THE NUMBER BIAS

THE NUMBER BIAS

HOW NUMBERS LEAD AND MISLEAD US

Sanne Blauw

Translated from the Dutch by Suzanne Heukensfeldt Jansen

SCEPTRE

Originally published in Dutch in 2018 by De Correspondent,
Amsterdam, Netherlands as *Het bestverkochte boek ooit (met deze titel)*
First published in Great Britain in 2020 by Sceptre
An Imprint of Hodder & Stoughton
An Hachette UK company

3

Copyright © Sanne Blauw 2018 and 2020
Dutch translation © Suzanne Heukensfeldt Jansen 2020

The right of Sanne Blauw to be identified as the Author of the Work has been asserted by her in accordance with the Copyright, Designs and Patents Act 1988.

This publication has been made possible with financial support from the Dutch Foundation for Literature.

All rights reserved. No part of this publication may be reproduced, stored in a retrieval system, or transmitted, in any form or by any means without the prior written permission of the publisher, nor be otherwise circulated in any form of binding or cover other than that in which it is published and without a similar condition being imposed on the subsequent purchaser.

A CIP catalogue record for this title is available from the British Library

Hardback ISBN 9781529342734
Trade Paperback ISBN 9781529342741
eBook ISBN 9781529342758

Typeset in Adobe Jenson Pro by Palimpsest Book Production Ltd,
Falkirk, Stirlingshire

Printed and bound in Great Britain by Clays Ltd, Elcograf S.p.A.

Hodder & Stoughton policy is to use papers that are natural, renewable and recyclable products and made from wood grown in sustainable forests. The logging and manufacturing processes are expected to conform to the environmental regulations of the country of origin.

Hodder & Stoughton Ltd
Carmelite House
50 Victoria Embankment
London EC4Y 0DZ

www.sceptrebooks.co.uk

Nederlands
letterenfonds
dutch foundation
for literature

For my mother.

For my mothers

CONTENTS

Foreword: Captivated by numbers   1

1. Numbers can save lives   9
2. The dumb discussion about IQ and skin colour   25
3. What a shady sex study says about sampling   49
4. Smoking causes lung cancer (but storks do not deliver babies)   72
5. We should not be too fixated on numbers in the future   99
6. Our psychology decides the value of numbers   120

Afterword: Putting numbers back where they belong   131

Checklist: What to do when you encounter a number   135

Notes   139

Further reading   161

Acknowledgements   163

FOREWORD

CAPTIVATED BY NUMBERS

She entered the dusty office through the sliding door and shook my hand. 'Juanita.'[1] In her oversized, faded jumper she looked even smaller than she was. Once she had seated herself on the folding chair opposite me, I explained to her in Spanish that I was doing research in Bolivia into happiness and income inequality for a Dutch university. I said I wanted to ask her a few questions about her own life and her country.

This wasn't the first time I had done this little spiel. For ten days I had been interviewing the inhabitants of Tarija, a Bolivian town close to the Argentine border. I had spoken to female market traders, drunk beer with strawberry farmers, barbecued with families – all to collect as much data as possible. Now I had arrived with my stack of questionnaires at the office of a women's organisation, whose director had offered to put me in touch with *empleadas domésticas*, female domestic workers. Women like Juanita.

'Let's start,' I said. 'How old are you?'

'Fifty-eight.'

'Which ethnic group do you belong to?'

'Aymara.'

Aha, I thought, she belongs to one of the indigenous populations. This was something I hadn't yet encountered.

'What's your marital status?'

'Single.'

'Can you read?'
'No.'
'Write?'
'No.'

My questions continued along these lines – her profession, her education, whether she had a mobile phone, refrigerator or television.

'I earn two hundred boliviano a month,' she told me when I asked what her salary was. This was far below the minimum wage of 815 boliviano that President Evo Morales had introduced not long before. 'I'm afraid that my boss will sack me if I ask for more money. I live in a *carpita*.' I noted down the word but did not know what it meant until I looked it up afterwards. She lived in a tent.

Eventually I got to the crux of my research: happiness and income inequality. Behind my desk on the eleventh floor of Erasmus University in Rotterdam, I had created five PowerPoint diagrams. Each one represented a different income distribution.

Only one day into the start of my research project in Bolivia I had noticed that my question about income inequality did not work for everyone. The market traders I had interviewed did not understand what the diagrams were meant to represent. How could I expect Juanita – who was not able to read or write – to understand this question? I decided to skip this part.

But before I could continue, she began to talk. 'Do you know what the problem is with Bolivia?' She sat up straight. 'There is a very big group of poor people and a very small group of rich people. And these differences are only getting bigger. Are you surprised no one in this country trusts anyone any longer?'

Without knowing, she had described diagram A. And in the process, she had answered two of my other questions as well, about her outlook on the future and trust in her country. I had totally underestimated her. My face flushed, but I continued with the interview as if nothing had happened. Time for the last questions.

'On a scale of one to ten, how happy are you?'
'One.'
'And how happy do you think you'll be in five years' time?'
'One.'

I think that it was during this interview in 2012 that I began to develop misgivings about numbers. Until then, I had been primarily a consumer of numbers. I came across them reading the paper or watching the news. For assignments during my degree course in Econometrics, I had been handed files with numbers by my professors or I had downloaded official data from the websites of organisations such as the World Bank.

But this time I was not given a ready-made spreadsheet. Now I was the person collecting the data. I was a year into my PhD; numbers had become my expertise. But my conversation with Juanita made my faith waver. I was researching her happiness but had found there was no number to express her life spent living in a tent. I listened to her opinion about income inequality, but was only able to choose from diagram A, B, C, D or E. Much of what she told me could not be counted, but did count.

Juanita taught me something else. I exerted a strong influence on what the figures looked like. *I* had decided that happiness was important and that it could be quantified. *I* had come up with the idea to ask this abstract question using diagrams. *I* thought that Juanita was not intelligent enough to have something to say about income inequality. I, I, I. Someone else, with the same research questions, but with a different worldview or angle, would probably have come up with different results. Numbers were supposed to be objective, but I suddenly saw how strongly they were linked to the researcher.

After my chat with Juanita, I typed her data into row 80 of my Excel sheet: 58 for age, 200 for salary, 1 for happiness. It looked as neat and tidy as all the other spreadsheets I had downloaded over the years. But it suddenly struck me how misleading my orderly rows and columns of numbers were.

I was already a number nerd as a toddler. As soon as I learnt to count, I devoured join-the-dots books. One of my first memories is of a holiday in the Black Forest in Germany, during which I spent countless hours tracing numbers to create a never-ending succession of snowmen and clouds. A few years later, my grandparents gave me a radio alarm clock. At night, I would lie in bed staring at the LED-illuminated clock and formulate all sorts of sums from those four digits. Maths was my favourite subject at school, and I would ultimately go on to study and get a PhD in Econometrics. I learnt everything about the statistics behind economic models. I calculated, analysed, programmed. And so I found myself again doing what I had done in my time connecting the dots: finding patterns.

But numbers also played another role in my life. They helped me to find my place. Between the ages of five and twenty-six I was awarded marks and grades at school and university. I used them as a yardstick for how I was doing. If I got a low mark, I'd be down in the dumps. A high mark and I was flying. It did not bother me that I'd forgotten the material within days, as long as I had a decent overall average. Outside school, too, numbers grounded me. When I returned from Bolivia I checked myself on the scales: 56 kilograms. I knew that would make a BMI of 18.3 – I was so proud.

I was not the only person taking my cue from numbers. Colleagues at university were promoted if they had published lots of papers in scientific journals. In the hospital where my mother worked, they awaited with trepidation the annual Top 100 Hospitals ranking. And my father had to retire the day he turned sixty-five.

It only struck me later that my exchange with Juanita had revealed something significant about these kinds of numbers. Just as I had shaped the numbers I had collected, so others influenced the numbers people used all around me as a guide for their lives. Teachers would calculate the correct mark for an exam, doctors the optimum BMI level, policy-makers the age at which you should stop working.

After I finished my PhD in 2014 I decided to go into journalism, because I had learnt something else from my conversation with Juanita: I found the stories behind the numbers even more interesting than the numbers themselves. At the *Correspondent*, an online journalistic platform, I started working as the numeracy correspondent. Not only did I want to explain to readers how numbers are calculated, I also wanted to question their importance in our society. Should we not put a stop to the dominance of numbers?

It soon became clear that my idea had struck a chord. Readers sent me skewed polls, shaky scientific research, misleading graphs. Many of the errors were ones I had made during my PhD research. At conference talks and in reviews of my articles, it became clear to me that my samples had not been representative and that I had mixed up correlation and causation. Now I saw the exact same mistakes appear in the numbers journalists used to interpret the world, Members of Parliament to make policy choices, and doctors to make decisions about our health. The world proved to be rife with false numbers.

Other kinds of reports about numbers also troubled me. I heard about parents whose nursery would hand them a school report for their one-year-old child, police officers who issued fines to meet quotas, Uber drivers who were sent packing when their ratings were too low.

It became increasingly clear to me that – from pension age to Facebook clicks, from GDP to salary – numbers determine how the world works. And the power of numbers only appears to be increasing. Big data algorithms are mushrooming in the public and private sectors. More and more, it is not people, but mathematical models that call the shots.

It's as if we have been hypnotised en masse by numbers. Whereas words are picked apart at the drop of a hat, numbers are given considerable free rein. After a few years as a journalist I have come to the conclusion that numbers have far too much influence in our lives. They have become so powerful that we can no longer ignore their misuse. It's time to end the dominance of numbers.

But, don't get me wrong, this is not a book against numbers. Like words, numbers are innocent. It's the people behind the numbers who make the mistakes. This book is about them, about their mistakes in reasoning, their gut feelings, their interests. We will be meeting psychologists who couch their racism in statistics, a world-famous sexologist with a shady data collection process, and tobacco magnates who massage their figures and ruin millions of lives as a result.

This book is also about us: number consumers. Because we allow ourselves to be led and misled by numbers. Numbers influence what you drink, what you eat, where you work, how much you earn, where you live, who you marry, who you vote for, whether you get a mortgage, how you pay for your insurance. They even influence whether you fall ill or recover, whether you live or die.

You have no choice; even if you're not a numbers person, numbers rule your life.

This book sets out to demystify the world of numbers, so that *everyone* can distinguish when they are being used correctly or when they are being misused. And so that we can all ask: what role do we want numbers to play in our lives?

It's time to put numbers in their place. Not on a pedestal, not out with the rubbish, but where they belong: alongside words.

Before we reach that point, though, we need to go back to the beginning. Where and when did our obsession with numbers begin? Allow me introduce you to the most famous nurse in history: Florence Nightingale.

CHAPTER 1

NUMBERS CAN SAVE LIVES

She would never forget the living skeletons.[1] The British soldiers languishing on rotten wooden camp beds, vermin crawling all over them. They died, one after another.

Slaughterhouses, that is what they were, the overcrowded hospitals in which Florence Nightingale worked during the Crimean War, the war between Russia and Britain, France, Sardinia and Turkey. Since the end of 1854, Nightingale had been stationed as the Nursing Superintendent of the military hospital in Scutari, to the east of what is now called Istanbul. But British military health care was so badly organised that she had to do much more than just nursing: cooking, washing, requisitioning for the stores. On some days she would work for twenty hours. After a few weeks she cut off her thick brown tresses because she did not have time for long hair. Her black dresses gradually became dirtier; a hole appeared in her white bonnet. If she managed to eat at all, in between mouthfuls she wrote letters to the outside world. Everything to keep her soldiers alive.

It was not enough; too many lives slipped through her fingers. 'We bury every twenty-four hours', she wrote in one of her many desperate letters to Sidney Herbert, the British Secretary of War. During the worst month, February 1855, more than half of the soldiers that were brought in died. Most did not die from their wounds, but from diseases that could have been prevented. The drains were so badly blocked that the ground underneath the building had become one big cesspool; faeces flowed directly from the latrines into the water tanks. Something had to change.

Meanwhile in Britain, the government collapsed following criticism of the shoddy warfare in Crimea. New Prime Minister Henry John Temple decided on a different course of action. He set up a 'Sanitary Commission' to prevent so many soldiers from dying in Scutari. And so, on 4 March 1855, four months after Nightingale arrived in Scutari, help finally arrived.

The commission found the situation in the hospital 'murderous' and set to work. They cleared more than twenty-five dead animals (including a horse in an advanced state of decomposition that was blocking the water supply). They drilled holes in the hospital roofs for better ventilation, whitewashed the walls, removed rotting floors. Towards the end of the war, in 1856, the military hospital in Scutari had been transformed beyond recognition. It was clean, well-organised, and the mortality rate had been drastically reduced. Not only the royal commission, but Nightingale too, had played a decisive role in this metamorphosis. Without her lobbying, the commission would probably never have made it to Scutari. Upon arrival in Britain, she was greeted like a heroine, a 'guardian angel'.

And yet she thought she had failed. 'Oh my poor men who endured so patiently,' she wrote in her diary after she left, 'I feel I have been such a bad mother to you, to come home and leave you lying in your Crimean graves.'

She was haunted by the needless deaths, the overcrowded wards, the vermin. The situation in the Scutari hospital may have improved, but care of the sick and wounded soldiers in the army was still organised in a woefully inadequate way. This cost lives.

Nightingale decided to fight for reform. She would use her experiences, network and newly acquired star-status to convince the powers that be of the dire need for better hygiene. And in her battle she would use a razor-sharp weapon: numbers.

The origins of our mania for numbers

Florence Nightingale was born in 1820 and grew up in a well-to-do British family. Her father was a progressive man: he believed girls were as deserving of an education as boys. So Florence and her sister Parthenope – both named after the places where they were born – were taught physics, Italian, philosophy and chemistry.

Florence was also taught mathematics, a subject in which she excelled. From an early age, she had harboured a fascination for counting and categorising. She started to write letters from the age of seven, in which she would often include lists and tables. And she was a great fan of puzzle books with riddles such as: 'If there are six hundred millions of Heathens in the world, how many Missionaries are needed to supply one to every twenty thousand?'

She would never lose her interest in numbers. When the then Minister of Defence asked her in 1856 what the situation was like in Crimea, she seized her opportunity. Over a period of two years she wrote an 850-page report in which she made use of numbers to show what was wrong with medical care in the army.[2] Her most important conclusion: many soldiers died from preventable causes such as wound infections and contagious diseases. Even in peacetime, British soldiers – who were being nursed in military hospitals – died in greater numbers than sick civilians. *Twice* as many. No less criminal, Nightingale thought, 'as it would be to take 1100 men per annum out upon Salisbury Plain and shoot them'.

As shocking as this conclusion may have been, Nightingale feared that it would get lost in the hundreds of pages of words and statistics. So she decided to cast her statistics in colourful graphics that would convey her point at a glance. Her most famous graphic shows two diagrams representing the two years of the Crimean War. Nightingale shows what soldiers had died from each month. Time and again most men died from diseases that could have been prevented.

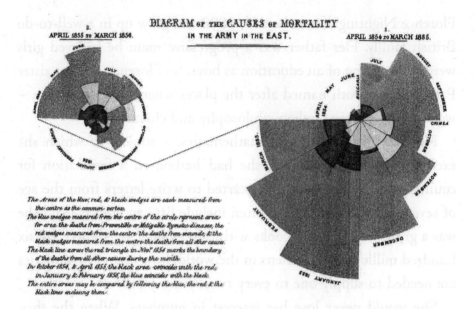

'Diagram of the Causes of Death or Mortality of the Army in the East', the graphic that Florence Nightingale published in her hefty report about medical care in the British Army.
Source: *Notes on Matters Affecting the Health, Efficiency, and Hospital Administration of the British Army* (1858).

She sent these and other charts to influential people, such as former Secretary of State Sidney Herbert, who by then was presiding over the Royal Commission on the Health of the Army. She also leaked her findings to the press,[3] and asked the writer Harriet Martineau to write an article for the wider public about the need for reforms.[4]

Nightingale was ultimately able to convince the authorities with her figures. During the 1880s, many problems had been solved: soldiers were better fed, had more opportunities to wash themselves, and their barracks were cleaner.[5] The situation improved so much that the newly built hospitals soon proved too big. 'Really it is not our *fault* if the number of sick has fallen so much that they [the Army Medical Department] can't fill their hospitals,' Nightingale remarked wryly.[6]

Florence Nightingale was one of the first people in the world to use graphs to effect change.[7] There is no doubt that she was intelligent, hard-working, stubborn, but she was also a product of the particular times in which she lived. During the nineteenth century, for the first time in history, statistics were being used extensively, a development that has continued to this day. The century saw the advent of nation states, which, with their growing bureaucracies, required more information from their citizens. Who died, who was born, who married whom – this was the first time this kind of information was recorded on a large scale.[8] 'An avalanche of printed numbers', philosopher Ian Hacking called this development.[9] Technology researcher Meg Leta Ambrose referred to it as 'this first wave of big data'.[10]

Figures on poverty and crime, our civil registration offices, and the averages and charts you see in the paper every day, all have their roots in the nineteenth century, less than two hundred years ago.

All this did not spring out of nowhere. To understand why Nightingale and her contemporaries started using figures on a large scale (and were able to do so), we need to delve more deeply into history, into three important developments that preceded the nineteenth-century mania for numbers.

We begin to standardise

We have been counting since time immemorial.[11] The oldest written messages handed down contain symbols that refer to numbers. A clay tablet from Uruk, a former city in present-day Iraq, records

'29,086 measures barley 37 months Kushim', some 2,400 to 3,000 years BCE. The text probably means that a certain Kushim received almost 30,000 measures of barley over a period of thirty-seven months.

He may have been the first person whose name we know, historian Yuval Noah Harari writes. 'It is telling that the first recorded name in history belongs to an accountant, rather than a prophet, a poet or a great conqueror.' It is most certainly telling, because numbers were crucial for the development of a society.

As a hunter-gatherer, you would have been able to remember all the information you needed. Where prey was lurking; which berries were poisonous; who you could trust. As a farmer in a small community you would store all the necessary knowledge in your head. But following the agricultural revolution, people began to cooperate on an ever-larger scale, in towns and cities and even in countries. The economy became increasingly more complex; money was introduced instead of barter, and a more and more opaque network of economic relations developed. You were in debt with one person, were owed money by someone else, had to pay rent to a third. And so our species hit a barrier; we were no longer able to remember everything.

This was a particular issue in states that wanted to levy taxes from large numbers of people. An official needed a method to record all incoming and outgoing payments. That method became script. By writing down agreements – legislation – and keeping a tally of who had done what – administration – it was no longer necessary to remember information. And much that was written down, as in the case of Kushim's barley, contained numbers.

The first development of numbers revolved not only around the sheer fact *that* we began to record them, but also around *what* we recorded. Let's go back to a specific part of Kushim's message: '29,086 measures'.

In his case you not only had to agree about the figure, 29,086, but also about what 'a measure' meant.

For the best part of history, agreements about measures were very local.[12] Every place used its own measuring unit that suited that particular location. In France, for instance, land was measured in *bicherées* – the number of *bichets* ('bushels') of grain a farmer needed to sow that field; or *journaliers* – the area of land a grape-picker could cover in a day.[13] (In the English language there are still traces of these old-fashioned measures: a stone's throw, within earshot.) Even if different regions used the same measure, its precise meaning could vary widely. In the eighteenth century, for example, 'a pint' in Précy-sous-Thil in France was more than three times the size of 'a pint' in Paris, 200 kilometres away.[14] It has been estimated that in eighteenth-century France, *a quarter of a million* different measurements for length and weight existed.[15]

In the same way that you cannot understand each other if you do not speak the same language, you cannot enter into agreements if you use figures in different ways.[16] In 1999 an incident took place that showed how dangerous it can be not to have a common number language. That year, the space probe Mars Climate Orbiter was due to reach Mars. But on 23 September 1999 the probe disappeared off the radar. The spacecraft was never found again. How could this have happened? To operate the probe, two computer programmes had to communicate with each other. One measured in 'pound-force seconds', as the American–British system dictated, while the other used the internationally accepted 'newton seconds'. The result of this miscommunication was that the probe flew 170 km lower than planned and was probably destroyed in Mars' atmosphere.[17]

Fortunately, such problems are the exception rather than the rule these days, because almost every country in the world now uses the International System of Units. But this shift did not happen without incident – it even required a revolution. After the French Revolution (1789–99), the revolutionaries decided to ditch all local units of measurement. They came up with a new proposal: the metric system. Units such as the metre and the kilogram dovetailed neatly with the ideas of the scientists of the time, plus – an important consideration – they would make the country more governable.[18]

How can you levy taxes as a state, if everyone uses a different measure for distance? You can't, of course, but a solution was found. It took a while, but eventually the metric system – later the International System of Units – would spread from France to almost all the countries in the world. Just three countries – United States, Liberia and Myanmar – use different official measures, such as pounds and miles.[19]

This was the first development underpinning Nightingale's thinking: we began to standardise. In other words, we agreed on how we would measure a particular concept. The metre and the kilogram were just the beginning. In Nightingale's time, half a century later, there was a craving for more numbers. Migration from the countryside meant that cities were bursting at the seams, and all kinds of problems coalesced and became visible: poverty, criminality, disease.[20] Where did those problems come from? And how should they be tackled? More and more people began to question this, within and beyond government.

To be able to gauge the seriousness of the problems, clear categories had to be devised: when was someone poor, criminal or ill? For instance, William Farr, a famous statistician who helped Florence Nightingale with her report, together with his colleagues, came up with a list of recognised diseases, which would eventually be taken up by the World Health Organization (WHO). Nightingale also used categories in her charts when she showed how many men had died of (1) preventable diseases, (2) war wounds and (3) all other causes.

On the face of it, the definition of a concept such as 'disease' or 'cause of death' appears to bear no relation to figures, but nothing is further from the truth. Something can only be made quantifiable when a clear definition is used. In the words of the philosopher Ian Hacking: 'Counting is hungry for categories.'[21]

As a result of standardisation we ended up speaking the same number language. Today, throughout the world, people speak of metres and kilograms, GDP growth and IQ points, CO_2 emissions and gigabytes. And so the most widely spoken language in the world did not become Chinese, English or Spanish, but numbers.[22] And this number language enabled the next development: we began to collect numbers on a large scale.

We began to collect numbers

As we saw from Kushim's clay tablet, numbers have been collected and recorded for millennia. But in the case of Kushim, it was still only a small-scale measurement (historians think he may have been responsible for a storehouse for beer-making ingredients).[23] During subsequent millennia, authorities began to collect numbers on a larger scale. One of the most famous stories from Western culture, the birth of Jesus Christ, would never have happened in Bethlehem if the Romans had not wanted to know how many people lived in their empire. History is peppered with such censuses – from ancient Egypt to the Inca Empire, from Han China to medieval Europe.[24]

William the Conqueror went one step further in 1085. He wanted to register the property of everyone in England. The Domesday Book would include the data of more than 13,000 places in England and Wales. Each place was visited by a group of officials who noted down more than 10,000 facts per shire: the owner of the estate, the number of serfs working the estate, the number of mills and fish ponds, etc.[25] It's difficult to comprehend how time-consuming this exercise must have been.

For centuries, the scale of the data collection in the Domesday Book was exceptional. It was not until the nineteenth century that the amount of data available would grow exponentially.[26] It was a time when many organisations were set up to collect figures. This was often done by the state (the term statistic derives from the word 'state', after all). In 1836, the General Register Office for England and Wales was created, which was responsible for registering births and deaths, but soon began to conduct censuses.[27] Beyond the confines of government, organisations also began to compile figures. The British East India Company, for instance, recorded who was ill, who had died and who was no longer employed by the Company, for around 2,500 employees.[28]

Nightingale's wish to improve medical care in the army during the middle of the nineteenth century fitted the zeitgeist: all around her, figures were being collected. But it needed one final development to bring about real change. Compiling mountains of figures is one thing, making sense of them is another.

We begin to analyse numbers

These days, you can't open a newspaper without coming across a chart. But the concept of casting figures into images is relatively new. Bar and line charts were only invented by William Playfair at the end of the eighteenth century. Nightingale would later use his ideas to draw attention to the grievous situation in army medical care, because charts could explain a mountain of numbers in an instant.

When, at the beginning of the nineteenth century, more and more figures were being collated, the need to analyse them also increased. In conjunction with the graph, the 'average' became popular. Nightingale used this method extensively in her voluminous report, for example to calculate the average number of patients per month during the Crimean War.

Perfectly normal though 'average' may seem in this day and age, in Nightingale's time the concept was still new. That is to say, as far as data about people was concerned; the average had been used by astronomers since the end of the sixteenth century. What if I applied this to human beings, instead of celestial bodies, Adolphe Quetelet thought in the nineteenth century.[29] This Belgian astronomer was one of Florence Nightingale's idols; she called him the 'creator of statistics'.[30] In an earlier life, he had been the director of the Brussels Observatory, but his building had fallen into the hands of freedom fighters during the Belgian Revolution of 1830.[31] The incident made Quetelet wonder: why do people do what they do? On the face of it, society seemed to be a chaotic muddle; that much was clear from the situation in his own country. But it should be possible to find a pattern in human behaviour, he believed.

Quetelet came up with a ground-breaking notion: 'l'homme moyen', the average man.[32] He made frantic calculations of averages for length, weight, criminality, education, suicides... and he devised the Quetelet Index, now better known as the body mass index (BMI), a measure to gauge whether someone's weight was within 'normal' range. Doctors, insurers and dieticians still use this measure to assess whether someone has a healthy weight.

In the footsteps of charts and averages, increasingly complex methods for analysing numbers would follow as the nineteenth century was ending. Historian Stephen Stigler would designate the period between 1890 and 1940 'the Statistical Enlightenment'.[33] Scientists at the time invented ingenious ways to find patterns in numbers, such as working out correlations and designing experiments.

Florence Nightingale sadly would not live to see much of this, as she died in 1910. Yet her number crunching had been groundbreaking. Almost a century after the Crimean War, a Scottish physician would follow in her footsteps and show once more that you can save lives with numbers.

It was August 1941 and prisoner Archie Cochrane was poised to tell the Germans about his secret experiment.[34] The Scottish doctor must have presented a wild appearance with his big red beard and emaciated face. Below his khaki Bermuda shorts his knees bulged, full of fluid.

He was not the only soldier with swollen knees. One by one, his fellow prisoners of war in Salonica (Thessaloniki) in Greece also began to complain of oedema. Cochrane, appointed chief doctor of the camp by the Germans, counted twenty new cases every day. He even reported the figures a little lower than they actually were, keen not to alarm his fellow prisoners more than necessary. But now it was time to speak up. He had decided to ask the Germans for help to save the lives of his patients. Not that he expected much from them. Only recently, one of the sentries had thrown a hand grenade into the latrine because he had heard *'verdächtiges Lachen'*, suspicious laughter.

Cochrane had an inkling of what the cause of the fluid build-up might be: wet beriberi, a disease caused by vitamin B deficiency. He decided to follow the example of his hero James Lind nearly two centuries previously. In 1747, naval doctor Lind had conducted one of the first clinical trials in history. He had divided twelve sailors who were suffering from scurvy into groups of two, each with their own diet. One pair was given six spoons of vinegar daily, another 250 ml of seawater, a third oranges and a lemon, and so on.

Lind soon detected a pattern: the sailors on the citrus fruit diet were considerably improved within a few days. He had discovered what is now well known: that scurvy can be prevented if you consume enough vitamin C.[35]

In Salonica, Cochrane divided his twenty patients into two groups. One group was given, three times a day, a supplement of yeast, a source of vitamin B that he had managed to procure on the black market; to those in the second group he gave a vitamin C tablet from his emergency supplies.[36] Nobody knew about his experiment.

On the first morning he noted down how often the patients had passed water. There was no difference between the groups. On the second day, again, there was no difference. But then, on the third day, the urination numbers for the yeast ward were slightly higher. On the fourth day Cochrane was convinced: the men who had been given yeast retained less fluid and passed more water. What's more, eight out of the ten men said they felt better, while the other group was still in a wretched state.

Cochrane had recorded it all neatly and now stood in front of the Germans with his logbook. They had to do something to help, he pleaded. The consequences would be dire otherwise.[37] Quite to his surprise, the Germans seemed shaken by his story. A young German doctor asked him what he needed. 'A lot of yeast,' Cochrane replied, 'at once'. The following day a large supply of yeast arrived. Within a month, hardly any patients at the camp were suffering from oedema.

Gut feelings, fallacies, interests

The story of Cochrane's trial is about more than finding new methods to analyse numbers. It's about the persuasiveness of numbers. Cochrane had even succeeded in getting his enemy, the Germans, on his side. What is it about numbers that tends to make them more convincing than words? Another event from Cochrane's life might help explain.[38]

Once he was back in Britain after the war, Cochrane began to argue for more statistics-based medical research. Medical trials of the kind he had conducted in the prison camp were still a rarity at the time.

In the 1960s in the UK, a series of exorbitantly expensive coronary care units were set up. At the time it seemed a logical development. Patients with coronary problems needed to be carefully monitored to avoid heart failure. But Cochrane, an out-and-out sceptic, was not fully convinced of this approach. If you really wanted to know the added value of such a unit, he argued, you should conduct a clinical trial: send one random group of patients home and keep one group of patients in the coronary unit.

He was roundly criticised by the ethics committee in London who alleged he was playing with peoples' lives. Yet Cochrane managed to persuade the committee's chairman of the value of his research. When he returned to his hospital in Cardiff, though, his fellow doctors refused to cooperate with him on the trial. They would decide how to treat their patients, they insisted. It infuriated Cochrane; what arrogance to think that they knew what was best for their patients. Medicine was more 'eminence-based' than 'evidence-based'.[39] It was more about the reputation of the physician than the scientific basis of his actions.

Cochrane's fellow researcher in Bristol did manage to conduct the trial in his hospital. Six months later, they both went to the committee in London with the results. These showed that the coronary unit had performed slightly better, but the difference was statistically insignificant. Yet the committee members – which had tried to thwart Cochrane six months previously – were indignant when they saw the figures. 'Archie,' one of them responded, 'we always thought you were unethical. You must stop the trial at once.'

Cochrane patiently allowed them to finish chastising him. When they had finished, he apologised, and revealed that he had shown them the wrong results. He produced a report with the real results: the same numbers, but reversed. Patients who were sent home did slightly better than those on the coronary care unit. Wouldn't you say now, he suggested, that the coronary care units should be closed?

This story reveals the obstacles that Cochrane had to overcome as a researcher. First, there was an emotional barrier. To physicians, it simply felt better and safer to keep their patients in hospital. The committee then made a wrong inference when they interpreted the information in such a way that it fitted their convictions.[40] Finally, vested interests played a part, because the reputation of the committee members would get a mighty knock if opening the hugely expensive coronary units turned out to have been the wrong decision.

Numbers were successful in conquering these three obstacles – gut feelings, fallacies and interests. Where words are easily coloured by bias, numbers can give a neutral representation of reality. In short, numbers seem naturally objective. It's no surprise that they have become so dominant in our society.

In 1993, five years after Cochrane's death, the Cochrane Collaboration – now simply 'Cochrane' – was set up, a worldwide network of health professionals and statisticians. This collaboration collates scientific evidence for almost any research field in medical science. The Cochrane Reviews are today some of the most important sources for evidence-based medicine.

Cochrane's plea for greater use of statistics in medicine has saved many lives. Take the Cardiac Arrhythmia Suppression Trial (CAST), an experiment conducted in the 1980s. During that time, doctors gave medicine to prevent heart arrhythmia to patients who had suffered a cardiac arrest. It seemed so logical: extra heartbeats tended to be associated with sudden death, so these had to be suppressed. But CAST – a comprehensive study among 1,700 patients – showed that the chances of dying after taking the medicine were not lower, but in fact *higher*.[41]

Cochrane's story, like Nightingale's, shows numbers in their best light: they can save lives. But there is another reason why numbers are so important. They help to keep rulers in check. Not for nothing is history peppered with politicians who meddle with numbers. For years in Argentina, the government ordered inflation rates to be manipulated.[42] Boris Johnson has been rapped over the knuckles numerous times by statisticians about the use of erroneous figures related to Brexit.[43] And Stalin had a statistician killed because he said that the Soviet Union's population was smaller than Stalin claimed.[44] An independent statistical agency can prevent politicians from using numbers to serve their own interest – and in so doing with how truth is perceived.

But numbers have a flipside. They can improve lives, but they can also destroy them. The three instruments that have been most important for the use of numbers on a large scale – standardisation, collection and analysis – are by no means foolproof. Sometimes things go wrong, very wrong.

CHAPTER 2

THE DUMB DISCUSSION ABOUT IQ AND SKIN COLOUR

During the First World War, 1.75 million American recruits completed an intelligence test.[1] This exercise was the brainchild of Harvard psychologist Robert Yerkes. Psychology, he thought, had the potential to be as rigorous a science as physics. But this meant he and his fellow psychologists had to collect the data.

His idea was a logical corollary of the nineteenth-century mania for counting. Not only was this a time during which units of distance and weight were standardised, it was also one in which researchers devised measuring methods for more abstract issues such as criminality and poverty.

And now it was time for 'intelligence' to be measured as well. Together with fellow experts, Yerkes designed the first intelligence test that could be conducted on a large scale and so in 1917 a study was conducted of historic proportions. Right across the USA, recruits were given a bundle of papers with questions that were supposed to gauge their intelligence.

Once Yerkes had compiled the data and was able to analyse it, a dreadful picture of the soldiers emerged.[2] White American men had the mental age of a thirteen-year-old; immigrants from Eastern and Southern Europe scored even worse. And right at the bottom – with a mental age of 10.4 – hovered the black man.

'I'd much rather have seen that black people are hyper-intelligent' (I)

Few people these days know who Robert Yerkes was, but the IQ of black people continues to be a topic that invokes heated discussion. 'There is a difference in IQ between nations', blogger and libertarian Yernaz Ramautarsing stated in 2016 in an interview with the Dutch news website *Brandpunt+*.[3] 'I would rather have seen something else, that black people are hyper-intelligent [...] But that is not the case.' His statement unleashed a furore two years later when he announced that he was going to stand as a candidate for the local elections in Amsterdam.

Ramautarsing is far from the only person to utter such claims.[4] Since Yerkes' test, discussion about intelligence and skin colour has cropped up in every new generation. In 1969, educational psychologist Arthur Jensen caused international uproar when he argued that the IQ differences between black and white students were the result of genetic differences.[5] In 1994, political scientist Charles Murray and psychologist Richard Herrnstein published *The Bell Curve*, in which they argued that, on average, African Americans had a lower IQ than white Americans, and suggested that women with a lower IQ should be discouraged from procreating.[6]

In 2014, there was yet another controversy: *New York Times* journalist Nicholas Wade wrote the bestseller *A Troublesome Inheritance*. In this book he argues that the different 'races' were the result of evolution and that those differences manifest themselves in varying levels of intelligence and other features.[7]

Yerkes' test shows how far-reaching the consequences of such statements can be. Not that his study had been conducted particularly rigorously. Conducting an intelligence test among 1.75 million recruits may have seemed an impressive project; in reality, the figures were compiled hastily in a slapdash manner. In *The Mismeasure of Man*, Stephen Jay Gould describes how the rooms in which the recruits did the test had no furniture, were badly lit and were so crowded that, at the back, you could not hear what was being said. Some soldiers could not understand what was being said anyway as they had only just arrived in America. Others did speak English but could not read or write. The men, some of whom were holding a pencil for the first time in their lives, had to write down how many cubes they had counted or which symbol came next in a series.[8] And all this under time pressure, because the next group would often be waiting in the corridor.

Reason enough not to take the numbers too seriously, you might say. The opposite happened. Yerkes' conclusion that certain groups were less intelligent gave a scientific gloss to ideas that were already popular at the time. Eugenics, the science that purported to 'improve the human race', was all the rage in North America and Europe after the First World War. Time and again, Yerkes' figures were used in Congressional debates about American immigration policy. The groups of recruits that had performed so badly in this intelligence test – Southern and Eastern Europeans – had to be kept out, the politicians believed. Not much later, quotas were introduced for these groups,[9] which would keep millions of people on the other side of the American borders between 1924 and the Second World War.[10] Refugees who needed help, often Jews, were refused entry on the basis of these quotas.

Intelligence figures were also used to justify radical sterilisation laws. In 1927, it became legal in America to sterilise someone by force. 'Three generations of imbeciles are enough', the American Supreme Court declared. Only once tens of thousands of Americans had been sterilised was the practice outlawed in 1978.[11]

It's almost impossible not to react with indignation at this. But even if the consequences of an intelligence test can be abhorrent, it does not mean that the test results are flawed. Present-day tests show that Yerkes' conclusion still stands. On average, people with black skin have lower scores.

Does this mean that the statements about skin colour and IQ are correct? That Ramautarsing was right? *Absolutely not.* The discussion about IQ and skin colour is one of the ugliest examples of number misuse.

A few important caveats

Before we go on, what does it mean when someone claims that the IQ of one group is lower than that of another? First, statements about skin colour and IQ are often based on samples from the US. It is not the case, therefore, that *all* black people would have lower scores; only that black Americans in the sample scored lower than their white counterparts.

But there is a lot more to say about this. Statements about intelligence and skin colour always deal with an *average*: the average of one group is lower than that of another. Behind those two averages lie a whole range of scores, including African Americans who have high scores and white Americans who are positioned at the bottom of the spectrum. If you take the scores of the much-used Wechsler intelligence test, you see that the two groups overlap greatly (see figure). According to the test scores, many African Americans are more intelligent than the average white American. The reverse also applies: many white Americans have lower scores than the average African American. In short, these kinds of averages say very little about an individual.

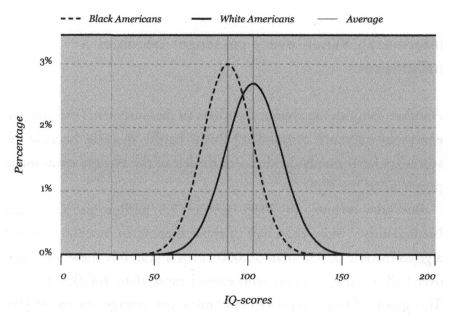

Scores on the Wechsler Adult Intelligence Scale (WAIS).
Source: William Dickens and James Flynn (2006)[12]

Another important question: what, in fact, is 'black' and 'white'? In studies, these labels are often based on how people themselves self-identify. But these categories are not cast in stone. Italians used to be considered non-white in the United States,[13] in Brazil you are black when you are non-European,[14] and in the 2010 census, millions of Americans entered a different category compared to 2000.[15] In other words, which category you belong to is determined as much by time and place as by your skin colour.

Even before you consider what IQ measures, these caveats – the origin of the data, the limitations of the average and the meaning of 'black' and 'white' – are important to bear in mind when dealing with hard conclusions about skin colour and intelligence.

Intermezzo: When every passenger becomes a millionaire

Another thing about averages: outliers in measurements can have an enormous influence. Around IQ this is hardly an issue, because the scores are fairly evenly divided – to the left of the average are as many people as to the right.[16]

But take income. In 2016, around 7.3 million people in the Netherlands – more than half of those earning an income – earned a gross income of less than 30,000 euros a year, but there were also over half a million people who earned more than 100,000 euros.[17] This group of high-earners sharply hikes the average. As an old joke among statisticians goes: when Bill Gates gets on a bus, on average, every passenger becomes a millionaire.

Because of the effect of outliers, you will sometimes hear about the 'modal' – or most common – income. 'Median' income is also used to avoid the effect of outliers. If you line up an entire population from low to high income, then the median income is that of the person in the middle.

Five subjective choices

It's time for the million-dollar question: what does IQ measure? Earlier we saw that standardisation, compilation and analysis were the most important developments for the widespread use of numbers. They are the same three steps researchers take when they work with numbers.

The first step – standardisation – plays an important role when we are talking about IQ. In order to standardise an abstract concept such as intelligence, researchers have to make choices along the way. Numbers may have an objective aura, but behind them tend to lie decisions that are subjective. Take the first scientists who engaged in IQ testing. They made five choices that were far from objective.

1. What you measure is a made-up entity

Robert Yerkes' test was inspired by the one devised by Alfred Binet, the founder of the IQ test.[18] This Frenchman would turn in his grave at the thought that the results of intelligence tests would be used to discriminate. In 1904, when Binet made intelligence measurable with the help of his student Theodore Simon, he had an entirely different aim in mind: helping children. The French Minister for Education had tasked him with developing a method to determine which children needed special education.

Initially Binet had tried to measure intelligence using a technique that had already been in use for some time – craniometry. The idea was that you could tell someone's intelligence from the size of their skull. But as Binet set to work with a measuring tape, he noticed that skull differences between high and low achieving pupils were *extrêmement petite*.

So, when the minister assigned him with this task, he decided to approach measuring intelligence differently. He created a test with exercises of increasing difficulty; the last question a pupil was able to answer correctly would indicate their 'mental age'. If this age was far below his or her true age, the child would require special needs education. Thus, Binet invented the first intelligence test. Following on his heels soon afterwards was psychologist William Stern with the famous intelligence quotient (IQ), which is worked out by dividing the mental age by the true age.

After the successful introduction of the universal kilogram and the metre, more things had been made measurable. In the case of distance and weight this was relatively easy, because everyone knew what the concepts represented: how far it was from here to there, how heavy a thing was when you lifted it. Such standards measured something *concrete*.

But, as we have already seen, from the nineteenth century onwards different types of numbers started to appear. Numbers about abstract concepts such as the economy, crime, education. Take the single concept that dominates everyone's life: money. Our coins and banknotes in themselves are worth nothing. You cannot eat them, you cannot use them to build anything, you cannot cure people with them.[19] But it has been mutually agreed that they are worth something. And we all rely on the fact that everyone – including the government – sticks to that agreement.

Such agreements have enabled us to cooperate on a much larger scale than would have been possible for hunter-gatherers. Nation states, human rights, religions – these are all human constructs for making sure we remain on the same page. But danger sets in when we begin to see such an agreement as objective. When we forget that we have come up with notions such as prosperity and education levels and think that these are set in stone. What happens then is called reification, from the Latin term *res* for 'thing'. Thingification, in other words. We come up with something, then forget that we came up with it and believe that it exists for real.

By measuring an abstract concept, it's accorded an even greater objective hue. Take the Gross Domestic Product (GDP), *the* measure for our economy. When our GDP falls, we are in a recession. If we then have to tighten our belts, it is because politicians believe it is good for GDP. So this particular measure has concrete consequences; you can lose your job, may have to pay higher taxes or may be eligible for financial support. GDP seems to work like an iron law of nature. Yet for all that, the concept is less than a hundred years old.

The idea of GDP was conceived in the United States in the years before the Second World War.[20] The country was stuck in a monumental depression. But the exact state of the economy? No one knew. There were some statistics about prices and transport, but not a single figure that summed up how the American economy was doing.

So the government asked the economist and statistician Simon Kuznets to measure the 'national income'.[21] Kuznets set to work; he methodically added up the income of households and companies. When he presented the first figures in 1934, the message was dramatic: between 1929 and 1932 the national income had *halved*.[22] For the first time ever, someone had taken the temperature of the American economy and it was well below zero.

During the years that followed, the American government became unhappy about Kuznets' 'national income' concept. With a war in sight, it proved politically awkward. The government preferred to spend money on arms rather than on people, but according to Kuznets' method, such government expenditure would have meant a drop in national income, and this, in turn, would have dampened support for the war. The solution was found in a different measure, the Gross Domestic Product. This would measure the total value of all goods and services produced in the country, *including* those generated by the government. From now on, new bombers were good for the economy.

Kuznets did not think much of this plan. He was convinced that a yardstick for the economy had to measure a country's *prosperity*. In his view, armaments had nothing to do with that. Kuznets lost the argument though, and, in 1942, the first American GDP was published – including defence expenditure.[23] It is clear that the resulting number had nothing to do with the laws of nature and everything to do with politics.

These days politicians and policy-makers tend to forget that GDP is an invented concept and use it as an objective measure, for instance when the government uses GDP to argue for cuts.[24] But GDP is not a concrete measurement like gravity. You do not make it any more 'real' by sticking a number onto it.

To come back to Yerkes and his soldiers' test: it is exactly the same with intelligence. It is an abstract concept, made up by people. A concept we began to measure.

Intermezzo: When three recessions suddenly vanish
Taking GDP seriously can be dangerous, especially when you forget that it is not always as precise as it seems.[25] In July 2015, the American Bureau of Economic Analysis announced that, during the previous quarter, the US economy had grown by 2.3 per cent. A month later this figure was adjusted to 3.7 per cent. And the month after that, it turned out to be 3.9 per cent.

Were the statisticians unequal to the task or in need of a holiday? No, adjusting economic figures is perfectly normal, and happens in any country collecting such measures structurally. This is not surprising when you see how much information is needed to work out this kind of figure. From taxes to defence expenses (yes, they are still included), from import to export – everything has to be thrown into the mix. Compiling such data takes time and will never be entirely successful. That's why it's so strange that the numbers are published in such precise detail – to one decimal place. (I will return to the uncertainty of numbers in Chapter 3.)

Supplementary data can at times present a radically different picture of the economy, for instance if a country is stuck in a recession. In 1996 economic data showed that the British economy had experienced ten recessions between 1955 and 1995 – periods in which there had been cuts and high unemployment, the entire country in disarray. But a newer dataset from 2012 showed a rather better picture; during the same period, the country's economy had been in recession only seven times. Three recessions had just vanished.[26]

2. What you measure is based on a value judgement

In 2007 researchers Shane Legg and Marcus Hunter – both AI specialists – collected all the definitions of intelligence they could find.[27] The crop was large; they found more than seventy different descriptions. Yet they saw some common ground and distilled all the various different definitions down to one sentence that was intended to encompass all of them: 'Intelligence measures an agent's ability to achieve goals in a wide range of environments'.

Legg and Hunter's suggestion may do justice to all definitions, yet it is still terribly vague. Within this framework, it is even considered intelligent to sneak unnoticed through the house at night to snatch a bottle of wine from the fridge. But you won't find such an exercise in an intelligence test.

What will you find? The Wechsler test includes exercises covering vocabulary, number sequencing and spatial skills – issues to do with abstract thinking.[28] This was already the case in Alfred Binet's initial intelligence test, which inspired Yerkes, in which children had to remember a sequence of numbers or state the differences between two objects.

To us, relating such abstract ideas to intelligence is self-evident. But a study from the early 1930s shows the limitations of this vision.

In his autobiography, Russian neuropsychologist Aleksander Luria describes his trip to Uzbekistan.[29] The country was modernising rapidly, and Luria wanted to see if these developments were leading to a different way of thinking. At some point, he and his colleagues visited Rakmat, a thirty-year-old farmer who lived in a remote part of the country.

They showed the man drawings of a hammer, saw, log and an axe. Which did not belong? 'They're all alike, I think all of them have to be here,' Rakmat replied. 'See, if you're going to saw, you need a saw, and if you have to split something, you need a hatchet. So they're all needed here.'

The researchers tried to explain to him that he had misunderstood the exercise. They gave the following example: let's say, you see three adults and a child; then the child does not belong. 'Oh, but the boy must stay with the others!' Rakmat replied. 'All three of them are working, you see, and if they have to keep running out to fetch things, they'll never get the job done, but the boy can do the running for them...'

The conversation with Rakmat shows that there are several ways to categorise, a standard part of an intelligence test. What if Rakmat thought up the questions for us? The test would probably measure whether we possess skills necessary for living within his community. The Uzbek would ask how you could best shoot a bird or how to pickle cabbage so that it would keep all winter. Most of us would fail spectacularly, as we would for a Masai or Inuit test. According to their standards, we are mentally challenged.

But it was not Rakmat who thought up our IQ test. Nor was it a nurse, carpenter or salesperson. It was people such as Binet and Yerkes: Western, highly educated men who had a fascination for numbers. How well you can care for someone who is ill, whether you can make a wooden table or whether you have social skills – in their tests these skills are unimportant. Completing number sequences, understanding metaphors and thinking in the correct categories: that's what their intelligence is all about. (This is exactly what I was expecting from my respondents during my research in Bolivia, and was the kind of thinking that I stupidly concluded Juanita could not manage.)

Abstract thinking has become so dominant in the meantime that it does seem as if this is the true form of intelligence. But there is nothing objective in deciding this type of thinking is the best. It is a *value judgement*.

The same goes for GDP. Simon Kuznets may have believed that this yardstick did not equate to prosperity, but since the Second

World War it has certainly been used as the defining measurement. To many governments, economic growth, a rise in GDP, is the supreme good. By choosing to treat it as such, a government makes a value judgement. Because what's part of the GDP is then by definition important. And yet it by no means always reflects what people value. A polluting industry, for example, is good for GDP, but bad for the environment. A less safe society means economic growth so long as people start putting in extra locks and security cameras.[30] And what to make of all the things that are *not* included in GDP? The Dutch, for example, spend twenty-two hours every week on care duties: cleaning, caring for others, looking after children.[31] This is not reflected in GDP. The irony is that if they had paid someone to do it for them, it *would* have featured in the GDP figures.

Not only are we measuring what we find important, it also works the other way around: what we measure *becomes* important. GDP is constantly used to underpin political decisions. Donald Trump, for instance, used economic growth as an argument for his trade war,[32] and a country's accession to the euro is highly dependent on its GDP.[33]

Likewise, performance on IQ tests has major consequences. They are frequently used in recruitment and selections processes and, to this day, the abstract thinking in these tests is still central to standardised examinations such as GCSEs and A-levels, which play a decisive role in someone's future.[34] We are in the grip of yardsticks of our own making.

3. What you measure is what you can count

The question remains: what exactly is intelligence? As we have already seen, the many definitions are so vague that it is impossible to translate the concept directly into numbers. If you want to measure something, you need a razor-sharp demarcation. In 1904, statistician Charles Spearman devised a trick that would render a definition of intelligence redundant.³⁵ Since, why would you capture something in words if you can let the numbers speak for themselves?

Spearman looked at the test scores and saw that people who did well in one test tended to do well in others. Some structure must be lurking behind all these tests, he thought, but what is it? He began to calculate and concluded that all scores per person could be translated into a single number,³⁶ which he called the g-factor, deciding that it measured a person's general intelligence ('g' as in 'general'). Just like Yerkes, he longed to turn psychology into a form of physics. With this method, his dream seemed one step closer. The confident Spearman considered his work 'a Copernican revolution in point of view'.³⁷ He published his findings in an article with the bold title 'General Intelligence Objectively Measured and Determined'.³⁸

But had he set to work as objectively as the title suggested? Even if we accept that in intelligence tests we only measure abstract thinking and leave many other attributes aside, we still have a problem: the sole input for Spearman's method was numbers. He only included what could be counted. In so doing he excluded all kinds of things that also relate to abstract thinking; things that are difficult to quantify – the quality of an essay, the creativity of a solution – or things that simply take too long for scientists to observe – how quickly someone learns a new language, how someone responds to making a mistake.

The upshot is that an IQ test never measures intelligence directly, but instead does so indirectly. The test result is a proxy, an approximation. There's nothing wrong with that; an IQ score helps psychologists gain an insight into the strengths and weaknesses of an individual. But they look beyond that one number. They study the results on individual test components and compare the figures to their own observations.

Only when the IQ score becomes *synonymous* with intelligence should we be concerned. And this is exactly what happens in the discussions about intelligence and skin colour. The IQ score is seen as *the* truth, instead of an *approximation*. It is exactly what psycho-logist Edwin Boring argued in 1923: 'Intelligence is what the tests test'.[39]

In our society, numbers are constantly seen as synonymous for the complex reality they are supposed to approximate. Take your work. In almost every job you are judged on quantifiable things. How many hours you work, how many clients you bring in, how many patients you help. But sometimes the truly important things are difficult to quantify: how sustainable is your relationship with your clients, how much kindness you show in your care. It brings to mind an aphorism that, it is rumoured, graced a wall in Albert Einstein's office: 'Not everything that counts can be counted, and not everything that can be counted counts.'

As with the IQ test, there is nothing wrong with keeping a numerical tally of your work. The data gives you an insight into the work you are doing. It becomes problematic when quantity is being confused for quality; if everything else you do during your working week is ignored and the focus is short-sightedly on the numbers alone. In the Netherlands, for example, police forces were judged on the number of fines they had issued.[40] The upshot was that special 'fine days' were organised, on which police officers had to issue as many fines as possible. You would suddenly be fined for minor offences such as cycling without lights or not wearing a seatbelt, which were normally overlooked. Whether this approach actually made society safer was of secondary importance.

Similarly, when the New Labour government in the UK decided that people should be seen in A&E within four hours, there was widespread manipulation of the targets by hospitals. People were kept in ambulances for longer and checked-in post-haste to meet the deadline.[41] According to the numbers, quality had improved, but the reality told a more sombre story.

Whereas the number of fines and the waiting times in A&E were once perhaps an adequate approximation of the quality of a police force or hospital, the figures soon became unreliable. The focus was no longer on things that were considered important, but on their *approximation*.

Time and again in such cases, you see people trying to find ways to manipulate the numbers. They adjust their behaviour or even commit fraud. This is sometimes called Goodhart's Law, after the economist Charles Goodhart. 'When a measure becomes a target, it ceases to be a good measure.'[42] Numbers are like soap: if you squeeze them too hard, they slip from your fingers.

4. What you measure will eventually be captured in one figure

One further important choice underpins the IQ score, namely that intelligence is to be captured in a single figure. Binet, the man behind the first IQ test, begged to disagree. 'The scale, properly speaking, does not permit the measure of the intelligence, because intellectual qualities are not superposable [. . .]'[43]

Over the years, many psychologists agreed with Binet. British-American psychologist Raymond Cattell spoke of two types of intelligence. On the one hand there is knowledge and experience – crystallised intelligence – and, on the other, there are skills such as logical thinking – fluid intelligence. Cattell was one of the architects of the Cattell–Horn–Carroll theory, which starts from the idea that there are several forms of intelligence, i.e. 'broad abilities' such as knowledge and pattern recognition.[44]

Nevertheless, despite all these different abilities, this theory also posits that intelligence can be captured in one all-encompassing g-factor. The theory has influenced many modern intelligence tests, which tend to calculate scores per ability, but they ultimately end up with one result: the IQ number.

Even Binet, who so firmly believed that intelligence could not be captured in a single number, came up with one figure per test person in the end: the mental age. Why? I have not been able to discover the exact reason, but I have a strong suspicion: it was neat and orderly.

When the economist Simon Kuznets first published his figures about the United States, the power of being able to summarise the national economy in a single figure was clear.[45] Whereas, before, all kinds of separate figures were available, now you could see at a glance how the wind was blowing. And this got people talking. Kuznets' published report became a bestseller – during an economic crisis, of all times – and President Franklin D. Roosevelt used Kuznets' figures as an argument for his programme that was to lift America out of the Depression.

In order to capture something as complex as the economy in a single figure, you will always have to leave something out. In the case of GDP figures, this is everything that cannot be expressed in terms of money. But the economist and philosopher Amartya Sen, who was awarded the Nobel Prize in 1998, argued that the development of a country is about more than money.[46] People should have access to good education and reliable health care, among other things.

It was this thinking which led him, in 1990, together with Mahbub ul Haq, to devise the Human Development Index, which today is a popular measure to gauge a country's development. This index looks at three factors: life expectancy, number of years of education and income. The higher the number, the more developed a country. In 2018, Norway came out on top with 0.95;[47] Niger came last with 0.38. The United Kingdom was ranked fifteenth.

Even though it is a good idea to use several factors to measure a country's development, yet again a complex concept is reduced to a single number. A number that can be communicated effortlessly. If you only have one number per country, it's easy to rank winners and losers, just as it's easy to rank people if you have a single figure for intelligence.

Intermezzo: When rankings are not actually rankings

The Dutch title of this book translates as *The Biggest Bestseller Ever (with this Title)*. The phrase is a nod to the rankings you see popping up everywhere: which country is the happiest, which doughnut the tastiest, which hospital the best – everything is numbered and ranked. Some of these rankings are utter nonsense. When an 'oliebollen' chef – a baker of the traditional Dutch doughnut – went on a Dutch TV talk show to complain that he had been awarded a 'one' in a newspaper's ranking – the lowest grade possible – it turned out that the figures had been tinkered with.[48] The jury apparently never gave a rating lower than a three. 'At our request, these figures were reworked into a scale of one to ten', editor-in-chief Hans Nijenhuis later admitted. 'That way there would be more contrast within the results.'[49] The Dutch newspaper in question, the *AD* (*Algemeen Dagblad*), has now stopped doing such taste tests.

The annual *AD* hospital ranking likewise says very little. Every year the paper selects a number of features on which hospitals are judged. In 2014 Dutch business expert Herm Joosten showed that hospitals rise or fall no fewer than twenty-five places on average.[50] Out of the hospitals that were in the top ten that one year, most had vanished into the lower rankings the following year. If you choose to go to the 'best' hospital, there is a strong chance that the hospital will no longer be the best by the time you find yourself in the operating theatre.

Back to using a single figure as an ultimate score for something as abstract as intelligence. There's another drawback: there are usually many different ways to measure the same concept. Take the Human

Development Index again. How do you add up life expectancy, education and income? What do you do with inequality in a country? And the differences between men and women: are those not also important factors to consider? These are all questions that have no single unambiguous answer.

I did not in fact make up these questions myself; in their report, the United Nations publish an inequality-HDI and a gender-HDI alongside the Human Development Index. They show how every country scores in the different areas, what the limitations of the measure are and the non-measurable dimensions.[51]

But such nuances rarely make it into the newspapers. Where one figure seems to give an easy insight, more figures throw a spanner in the works. You soon end up in a world full of ifs and buts. The figures about hunger, for instance, depend largely on how you define hunger.[52] The Food and Agriculture Organization (FAO) describes a person as undernourished when he or she does not consume a sufficient number of calories on a regular basis. But what is 'sufficient'? This can differ significantly between someone who spends his days typing behind a desk and someone who ploughs a field by hand.

The FAO itself made an alternative calculation in 2012, which showed that how you define hunger could totally alter the figures.[53] In one case, global hunger went up throughout the years; in the other, it went down. Then researchers also had to choose between the absolute number of people suffering from hunger and the percentage of the global population. An absolute number makes sense when you prioritise that every person counts. But a percentage is useful if you think it's important that the majority of the population gets enough nutrition. These are moral, not statistical considerations.

In IQ tests, likewise, research choices make a big difference to the test results. In 1984 psychologist James Flynn studied the numbers for previous generations and came to a surprising conclusion: their

IQ had gone up and up during the past century. If you calculate the scores of our forebears from the 1930s using the current test standards, with a score of 70 they turn out to be borderline mentally disabled. If you use their standards for the present generation, we end up with an average IQ of 130: highly gifted.[54]

Flynn discovered the effect eighty years after Alfred Binet conducted his first test with French pupils. Why had it taken so long for anyone to spot these enormous differences between generations?[55] Flynn's conclusion has been confirmed scientifically time and again since, but his measured effect is not something that can be seen with the naked eye. That's because, once in a while, the test gets updated.

The Wechsler test for children, for example, was used for the first time in 1949 and subsequently revised four times – in 1974, 1991, 2003 and 2014. Not only were the questions dusted down during these overhauls; so were the scores. The new test is tried out on a group of people and the IQ scores are calculated in such a way that the average of the test group works out at 100. The test groups – just like society – have been achieving higher and higher scores. Flynn found that we are now better trained in a particular kind of abstract thinking, which has become increasingly dominant in more and more schools and workplaces over the past century. If you have the exact same cognitive abilities as your ancestors, your IQ will nowadays end up lower than theirs.[56]

5. What you measure is what you want it to be

Let's return to Yerkes and his intelligence test of American recruits during the First World War. His team found that, according to the test results, immigrants were inherently unintelligent and that black men were at the very bottom of the ladder – but they also came up with a series of other results.[57] There turned out to be a strong correlation between a subject's test score and the number of years they'd spent in education.

Yet Yerkes did not conclude that education therefore led to a higher intelligence. He thought that the connection worked the other way around: 'The theory that native intelligence is one of the most important conditioning factors in continuance in school is certainly borne out by this accumulation of data.' When he noted likewise that black men had enjoyed less education, he did not see this as a reason for their lower score. It was their low innate intelligence that had led to their shorter time in education, Yerkes concluded, momentarily forgetting that they were living at the time of segregation.

Yerkes made exactly the kind of wrong inference that we will encounter extensively in Chapter 4: he accepted indiscriminately that a correlation was *causal*. That the colour of your skin determines how well you can think, even though his numbers were in no way able to corroborate his conclusion. He did not let the numbers talk, but instead trusted his gut feeling. And this gut feeling was in line with the times.

This is obvious from the preface that Yerkes wrote for *A Study of American Intelligence*, a book that was based on his dataset and was used frequently by eugenicists in the discussion about immigration in the United States. 'No one as a citizen can afford to ignore the menace of race deterioration or the evident relations of immigration to national progress', he wrote.[58]

You see this time and again, and we will encounter it frequently in this book: numbers are interpreted in a way that fits the beliefs or requirements of their users.

The inventor of the intelligence test, Alfred Binet, had already warned that we should not see intelligence as an immutable entity.[59] And yet Yerkes decided to interpret the numbers in such a way that the test scores did indeed stand for innate abilities.

Similarly, the man who put GDP on the map, the economist Simon Kuznets, warned that the figure did not equate to welfare.[60] And yet, during the course of the twentieth century, GDP has been used time and again to do just that.

These kinds of interpretations are dangerous. If you want to take figures seriously, you should acknowledge that there is a great deal that they do *not* say. In other words, that GDP is merely a measure for 'production' and IQ no more than a score on a test. Instead, due to convictions and biases, the figures are inflated into something they are not.

A century on, what can we say about Yerkes' interpretation of the soldiers' test scores? Do IQ figures really measure innate intelligence? They do not. As Binet suspected, our IQ is not cast in concrete. The most important evidence for this is the Flynn effect. The fact that IQ has risen over the generations does not mean that our ancestors were thick as two short planks and we are brilliant. We have simply got better at abstract thinking because it's a skill we're expected to apply so universally in modern life. In the words of Malcolm Gladwell: 'An IQ [. . .] measures not so much how smart we are as how *modern* we are.'[61]

Psychologists agree that our IQ is determined by both environment and genes. Living conditions can make an enormous difference. For example, in an IQ test taken before the harvest – a period of hunger and money problems – Indian farmers, on average, tended to score thirteen points lower compared to their score after the harvest.[62] Beforehand, their cognitive ability had been so consumed by their poverty that they had less space to think clearly.

In Kenya, another study revealed, the average IQ of children increased by more than twenty-six points between 1984 and 1998.[63] How could this be? The researchers pointed at improved conditions: parents were better educated, nutrition had improved and children were healthier.

Among African Americans, similarly, an improved environment led to higher scores. The difference in IQ with their white fellow citizens is smaller these days than in the past. Over the course of thirty years, African Americans closed the gap with white Americans by between 4 and 7 points.[64] In short, economist William Dickens and psychologist James Flynn (of the Flynn effect) concluded in 2006 it is a 'myth' that the IQ gap between black and white Americans remains the same.

To return to Yerkes and his followers, it was wrong to see IQ as synonymous with intelligence and it is complete nonsense to see it as *innate* intelligence. As long as the environment of black people is different to that of white people, it is futile to posit that the differences are caused by a fundamental biological difference between the two groups.

And even though there have been improvements, inequality between black and white is still very significant. In 2016 the median income of black families in America was 17,600 dollars, *a tenth* of the median income of white families, 171,000 dollars.[65] The schools in black, often poorer neighbourhoods tend to be of inferior quality to the schools in white neighourhoods.[66] And discrimination is still the order of the day. Experiments with fictitious CVs show over and over again that job applicants with African-American sounding names are rejected more often.[67] Being surprised that people score differently in a test is – I have no other word for it – idiotic.

'I'd much rather have seen that black people are hyper-intelligent' (2)

As we have seen in this chapter, a researcher will always have to make choices when he or she standardises an abstract concept like intelligence. Perhaps this makes it seem as if numbers serve no purpose. This is not the case. Numbers can help us detect patterns that would otherwise have remained hidden.

But it is dangerous to have the wrong expectations and to assume that, by definition, numbers are objective. Then numbers become an excuse to stop thinking. This is what happened when Yernaz Ramautarsing said: 'I'd much rather have seen [...] that black people are hyper-intelligent [...] But this isn't the case.' It's not my fault, he argued; it's what the numbers say.

This is an upside-down world. If we want to take numbers seriously, then we should recognise and identify all their limitations: that buried inside them are value judgements, that not everything can be counted, that there is so much that they do not say. That numbers are not *the* truth, but only an aid to understanding the truth.

Numbers can reveal things you would not have seen otherwise. We've seen, for example, how Archie Cochrane used numbers to test the effectiveness of drugs. IQ numbers, too, can be useful to help people. They give psychologists insight into the development of a child. And IQ scores that show a difference between black and white Americans can help us to grasp the depth of inequality.

Do not let a number be the end of a conversation, therefore, but a starting point. A reason to keep asking questions. Which choices have been made during the research? Where do the differences come from? How do they affect policy? And especially: does the number measure what we believe is important?

CHAPTER 3

WHAT A SHADY SEX STUDY SAYS ABOUT SAMPLING

A black-and-white photo from 1948 shows a middle-aged man holding up a newspaper with both hands. You can make out the headline in capital letters on the front page: 'DEWEY DEFEATS TRUMAN'. The man in the photo smiles so broadly it reveals a crack in one of his canines. He has just become the most powerful man on earth.

The photo is iconic, but not because presidential candidate Thomas E. Dewey in fact 'defeated Truman'. It is iconic because he did *not*. The man in the picture is, Dewey's opponent, Harry Truman.[1] And the paper in his hands was completely off the mark. Relying on opinion polls, the *Chicago Daily Tribune*'s editor-in-chief had been so convinced of Dewey's victory that he did not even await the results and printed the bold headline prematurely on election night.[2]

It could have been a photo of Donald Trump from November 2016. In his hands one of the many newspapers that had predicted Hillary Clinton's victory. On his face a broad smile, because they had got it wrong. 'How did he pull off such a stunning victory?' the *New York Times* asked the day after the election. 'How did almost no one – not the pundits, not the pollsters, not us in the media – see it coming?'[3]

Princeton professor Sam Wang had used opinion polls to predict that Clinton had a 99 per cent chance of winning. If Trump were to win, he had promised, he would eat an insect.[4] It tasted 'nutty', he said when he ate a cricket live on CNN, four days after the election.[5]

And so, almost seventy years after Truman's unexpected victory, the question of the reliability of opinion polls became relevant for the umpteenth time. Polls are not without consequence. They influence how the media write about politicians and who is allowed to take part in television debates. What's more, voters use polls when they want to vote strategically or to decide whether to head for the polling station in the first place. Polls thus directly and indirectly influence the election result. And with it our democracy.

The question of whether polls are reliable is about much more than elections. The method used to conduct the polls – sampling – is behind many of the numbers you encounter. It's data from a sample that is used when poverty is measured, when statistics about sexual harassment are collated, when drugs are tested. In these kinds of surveys, it's impossible to include everyone – all Americans, all women, all cancer patients. Physician Archie Cochrane did not study all the patients with oedema in the prison camp, just twenty of them. Psychologist Robert Yerkes did not test the intelligence of all American men, only soldiers.

The sample is thus the lens we use to understand the world.

The sample, Professor Jelke Bethlehem of Leiden University writes, is probably as old as humanity.[6] Everyone uses this method, consciously or not. For instance, when you're cooking you taste a spoonful of soup and judge the whole dish on the basis of that one sip. The Dutch term for sample, *steekproef*, has been used for centuries at Dutch cheese markets, where the tester 'sticks' (*steekt*) a cheese scoop into the cheese – to test (*proef*) – it.

It was in 1824, during the century that people really began to collect figures with a vengeance, that someone used a sample to poll opinions for the first time.[7] The American presidential elections that year were the most exciting since independence in 1776; not only was it a close race with four candidates fighting for victory, but also many Americans had only recently been given the franchise to vote.[8]

Voters were hungry for information and, entirely in keeping with the zeitgeist, people began to count. How often was a toast made to this candidate? Did people place bets on him? Soon, curious voters began to keep a tally of preferences during military parades, Independence Day parties or visits to a local bar. Newspapers published the numbers, especially if the results turned out well for their favourite candidate.

Let's wind the tape forwards to a good century later, when the broadly smiling Truman won the election in 1948. Polls had become more sophisticated in the meantime. They were conducted on a national scale by professional polling agencies and were no longer just about elections. From working women to the war, the United Nations to the common cold – Americans could now give their opinion on everything.[9]

But following the 1948 election, cracks appeared in the image of sample surveys.[10] If polling agencies had been so way off the mark over the election between Dewey and Truman, how could other samples be trusted? How reliable were their findings?

This emerging scepticism was directed at a controversial study that was also published in 1948. The book, stretching to 804 pages, covered a topic that made many people's eyes bulge: sex. It had been written by biologist Alfred Kinsey who, together with his colleagues Wardell Pomeroy and Clyde Martin, had interviewed 5,300 American men about their sex life.[11] *Sexual Behaviour in the Human Male* became a roaring success: more than 250,000 copies were sold and the book spent months in the national bestseller lists. There was barely a radio programme that did not feature it or a cartoonist who did not use it for a sketch.[12]

And everyone talked about the statistics in the report. The prevailing norms in the United States might have been virtuous but, according to the study, the reality was completely different. Ninety per cent of the men had slept with someone else before marriage, 50 per cent had been unfaithful and 37 per cent had had a sexual experience with another man. One in twelve men had had sex with an animal (one in six of the men who had grown up on a farm).[13] What is also striking is that the figures are still being used today. Have you ever heard that one in ten men is gay? It's from this study.[14]

But are these figures correct? The 1948 election fiasco showed that polls should be taken with a pinch of salt. *Life Today* magazine wrote, 'How much salt, then, must be taken with a poll which judges and condemns 60,000,000 white males on the basis of only 5,300 interviews?'[15]

The criticism snowballed and the Rockefeller Foundation, which had largely funded Kinsey's study, became restless. Finally in the autumn of 1950, three respectable statisticians set out to put the chief author of the sex report through the mill.[16]

Three statisticians go to a sex professor

The three renowned statisticians were waiting in a basement largely filled with books about sex. They really did not have time for this evaluation. Fred Mosteller had enough on his hands with his work at Harvard, William Cochran was chair of the Biostatistics department at Johns Hopkins University and, in addition to his responsibilities at Princeton, John Tukey was currently securing patent after patent for Bell Telephone Laboratories. It was out of a sense of duty that the three had made their way to the Institute for Sex Research in Indiana. Together, they were tasked with giving a definitive opinion about the quality of the much-talked-about sex survey.

They had barely arrived at the office temporarily assigned to them when the door flew open. There he was, with an army of secretaries and other staff members behind him. The man in charge of the institute that would host them, the man whose reputation depended on their judgement: Alfred C. Kinsey.

Professor Kinsey – Prok to friends – was a tall man who always wore a bow tie. His earlier research had been into the gall wasp. He had travelled across thirty-six American States and through Mexico to collect as many specimens as possible. Each wasp he had meticulously mounted, measured and recorded.

But in 1938 he was allocated a university subject that would rouse his interest in an entirely different field. He was given the opportunity to teach the Marriage and Family course at Indiana University. It was a course that was intended to prepare students for marriage; in other words: their sex life.

As a boy from an orthodox Christian family, Kinsey thought that there was something wrong with him when he was unable to stop masturbating. Sex was taboo at home, and he could not find any information about it. The only recourse was to pray to God to stop his sinful behaviour, the young Alfred concluded.

By the time he started to teach the marriage course, he was over forty and he knew better. But what was normal when it came to sexual behaviour? Still, no one knew. There was more data available about the gall wasp than about human sexuality. So he began to ask his students questions: Do you ever climax? Do you masturbate? Have you had sex with a prostitute? Yet Kinsey needed more data. He decided that he wanted to speak to 100,000 people for his dataset across the entire country.[17] He managed to convince the prestigious Rockefeller Foundation to fund his study. The foundation knew that sex was a touchy subject, but who better to research it than this happily married, somewhat nerdy professor? Kinsey would study people as if they were wasps, remaining neutral and detached. 'We are the recorders and reporters of facts,' he argued, 'not the judges of the behaviours we describe.'

In short: just facts, no opinions.

Two years after Kinsey's report had been published it was down to the three statisticians to assess whether he had done a good job. Their quest exposes six crucial mistakes that can be made when using samples.

1. The circumstances or questions are flawed

> 'What would you say was the main source of your early knowledge about sex?'
>
> 'Do you dream of giving or receiving pain, being forced to do something, or forcing someone to do something?'
>
> 'How old were you the first time you paid a female for intercourse or some other sexual activity?'

During their visit the three statisticians allowed themselves to be interrogated by Kinsey and his colleagues about their own sex lives. This meant they were able to experience first-hand how the interviews were conducted.

Kinsey's session lasted on average two hours and comprised – depending on the sexual experience of the test subject – 350 to 521 questions. The interviewer had learnt these questions by heart, the concern otherwise being that someone reading off a list would make the participants nervous. In order to guarantee confidentiality, the replies were noted in a secret, intricate code. ('P' could thus mean puberty, peers, petting or Protestant.[18]) In addition, Kinsey and his two co-interviewers tried to ask the questions in such a way that it was easy to share secrets. They did not ask: 'Have you ever cheated on your wife?' but 'During your marriage, how old were you *the first time* there was sexual intercourse with a woman other than your wife?'[19] John Tukey, the Princeton researcher, would have been surprised at such a question; he had only just got married to his wife, Elizabeth, whom he had met at folk dancing classes.[20]

The circumstances around an interview are crucial, especially when it deals with something as sensitive as sex. Practically every survey reveals that the number of sexual partners of the opposite sex is higher for men than for women. In a British study using data over a two-year period from 2010 to 2012, for instance, the average number of men that women said they had slept with was seven, while men mentioned on average twice as many female partners.[21] Yet this is impossible, because these extra women have to come from somewhere. Had the survey not been representative? Perhaps the men had more sexual encounters abroad? Or had they been going to sex workers, who had not been interviewed?

There was another plausible explanation: the subjects were not telling the truth. Take an experiment from 2003, in which 200 students were asked to fill in a questionnaire about their sex lives. Some were linked up to a lie detector. It was fake, but they did not know this. The result: the number of sexual partners for women shot up by 70 per cent – from 2.6 to 4.4 sexual partners.[22] This is just one of the many studies into lying in polls that demonstrate time and again how crucial the circumstances can be to the results.

And the circumstances around the Kinsey sex study? Were they the best they could possibly be? It's difficult to say. A comparative study shows that there is no single method that is best for research into sex. People sometimes turn out to be more honest when they have to complete a questionnaire on their own, but sometimes the interaction with an interviewer – as in the case with Kinsey – makes it easier to divulge sensitive information.[23]

In addition to the circumstances, the framing of the question is crucial in sample studies. Some questions, either by design or not, push respondents in a particular direction. Take Indian Prime Minister Narendra Modi's poll about a controversial policy measure. In November 2016 his government decided that the 500-rupee and 1,000-rupee notes in circulation at the time were no longer legal tender. People were given until the end of the year, barely two months, to exchange them.

Modi believed that the measure would combat corruption and tax avoidance. Moreover, it was intended to encourage the Indian public to switch to electronic payments, one of the Prime Minister's pet interests. But the decision met with huge popular protests. The measure was too radical, its opponents argued: 86 per cent of India's cash was at stake. Exchanging such an enormous quantity of money within just two months was bound to go wrong.

In order to silence the protests, Modi decided to conduct a survey. Within thirty hours, half a million people had answered his questions and the Prime Minister could be satisfied: more than 90 per cent found his plans good or even 'excellent'.

But look at the questions he asked:

- 'Do you think black money exists in India?'
- 'Do you think the evil of corruption and black money needs to be fought and eliminated?'
- 'What do you think of the Government's moves to tackle black money?'

- 'What do you think of the Modi Government's efforts against corruption?'
- 'What do you think of the Modi Government's move of banning old 500-hundred- and 1,000-rupee notes?'

In question after question the respondents were pushed towards the idea that this move was necessary to combat corruption. By asking questions to which it would be very hard to answer no – who doesn't think that 'evil' should be 'eliminated'? – you eventually end up at a point where it would be almost impossible to be *against* the initiative.

It reached true levels of absurdity when respondents had to say what they thought of the statement 'Demonetisation will bring real estate, higher education and health in the common man's reach.' There were just three options to choose from: completely agree, partly agree, can't say. Disagreeing was impossible: 'If you are in my marketing research class and design such a survey, I will fail you', Prithwiraj Mukherjee, a professor of marketing from Bangalore, wrote in response.[24]

A good survey asks neutral questions. This is easier said than done: even a subtle difference in the framing of a question can make a difference. In 2014, media company CNN and polling organisation Gallup simultaneously conducted a poll about terrorism.[25] Both were telephone polls, the groups were pretty much the same size and equally representative (more about representativeness later). And yet: 14 per cent thought terrorism was a big problem in the CNN poll, but just 4 per cent in the Gallup one. The difference was probably to do with how the questions were framed. CNN asked a closed question: 'Which of the following is the most important issue facing the country today?' Among the options – such as the economy and climate change – was terrorism. Gallup, on the other hand, asked an open question: 'What do you think is the most important problem facing this country today?' Without being prompted, people seemed less likely to think of terrorism.

Similarly, in Kinsey's sex survey there was a danger that the framing of the questions influenced the replies. He tried to encourage his respondents to speak the truth, but his questions could just as well have had the opposite effect. A question such as 'When was the first time you masturbated?' can make a masturbation virgin believe that he deviates from the norm, and therefore that it would be better to lie.

Yet, Kinsey's three inquisitors were impressed with their own interviews and considered it the optimal method for collating this kind of sensitive information. But their interview did not remove their concerns about the sex survey. They were not so much troubled by the questions or circumstances, as by something else entirely: the composition of the sample.

2. The survey excludes particular groups
The statisticians' big objection to Kinsey's study was that it was aimed at particular groups of people. Kinsey had collected data in gay bars, prisons and at universities. His methods were, to put it mildly, unconventional. 'We go with them to dinner, to concerts, to night clubs, to the theatre [...], in poolrooms, in taverns and get them to introduce their friends.'[26] Kinsey had even interviewed his own children. Over a period of some nine years, more than 11,000 people had talked about their sex life, around 5,300 men and a further 6,000 women, for a report that Kinsey would publish a few years later. And this with the assistance of only two colleagues, because these were the only people he trusted to do the actual interviewing. They worked long days and were continuously on the road.

However impressive this entire exercise may have been, sample surveying is not about quantity; it is about representativeness. And that was precisely the problem with Kinsey's method. There were many places he had not – or barely – visited: conservative church communities, factories, rural villages. Black men were entirely absent from his study.[27] Other groups – homosexuals, students, Midwesterners – were disproportionally well-represented. In short, a more appropriate title for the book might have been *Sexual Behavior in the Predominantly Midwestern White Human Male*.

Even today, it's often the case that only particular groups are approached for surveys. Take Modi's poll about his new measure. He disseminated his questionnaire via his own app, but, in 2016, only 30 per cent of the population of India had access to the internet.[28] The people who did have access came from a higher social class, tended to use bank cards instead of cash and generally had different political views from those who did not have mobile internet. What's more: if you're not in favour of the Prime Minister, the last thing you want is a Narendra Modi app on your phone. In addition, the questions were only posed in Hindi and English, which disenfranchised millions of people who did not speak either of these two languages.

Scientific research similarly often makes general statements while excluding certain groups. The field of psychology is dominated by research from Western countries, for instance. An overview paper from 2008 shows that as many as 95 per cent of studies from the previous five years were done with subjects from a Western country; the majority of these, 68 per cent, came from the United States.[29] Not only that, the subjects came from a very specific group: psychology students at research universities. These were close at hand and would often be happy to take part in a study just for a bag of M&Ms.

The samples in psychology are 'WEIRD', psychologist Joseph Henrich and his colleagues have argued: 'Western, Educated, Industrialized, Rich and Democratic'.[30] Research findings are often generalised to 'everyone', while WEIRD-people can differ greatly from other groups.

You can even see this in very basic psychological processes. Take the Müller-Lyer illusion whereby you are asked which line is longer, A or B (see the left-hand figure in the image). For most of us, line A looks longer. In reality, the lines are the same length, as you can see in the right-hand figure. It is a textbook example, but extra research among non-WEIRD communities shows that not everyone is susceptible to the illusion. A tribe in the Kalahari Desert, for example, saw no difference between the two lines.[31]

The Müller-Lyer Illusion

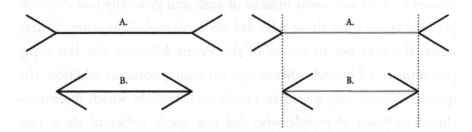

Excluding particular groups from a sample can have far-reaching consequences. Until 1990, most drugs were primarily tested on men,[32] as researchers did not want to run the risk of women being pregnant during the trial. The Thalidomide scandal in the 1950s and 60s – during which thousands of children were born with deformities because their pregnant mothers had taken the drug – had demonstrated how grave the consequences could be. In any case, it was thought that women were difficult to study, because their hormones fluctuate every month.

But women can react quite differently from men to certain drugs. When, in 2001, the American Government Accountability Office investigated medication that had been recalled because of harmful side-effects, it discovered that eight out of ten medicines affected women more strongly than men. Four of these drugs had been more frequently prescribed to women, but the other four were used equally by both sexes, and yet more women had suffered their side-effects. The drug Posicor, for instance, slowed down or stopped the heart in older women, but not in older men.[33]

Thankfully, over the past few years, action has been taken. Both the United States and the European Union have legislation in place that has resulted in a better representation of women in medical trials. But this does not change the fact that it can be critically dangerous to exclude certain groups from a sample.

3. The interview group is too small

The size of a sample does not guarantee that the study is representative. But the size of a sample group does matter. Take Archie Cochrane's research in the prison camp. He would later describe it as his most successful trial: assisted by the Germans, he had been able to combat oedema. But he also considered it to be his worst trial: he had only studied twenty men, ten in one group and ten in the other.[34]

The problem with a small sample is that it's much more likely to produce extreme results. Say you step outside and buttonhole the first person you meet. It turns out to be a woman. You talk to the next person and this passer-by also turns out to be a woman. It would be bizarre to conclude from this sample that 100 per cent of all people are female. The longer you carry on, the more people you talk to, the smaller the chance that the entire sample consists of women and the closer the sample comes to approximating the general population. That's why a survey with a small sample is never a good idea; your results may very well deviate strongly from the group you are interested in.

You see the same shortcoming in trials for which the samples are too small. When you compare two small study groups, there is a sizeable chance that one group will differ greatly from the other, because one outlier can easily create a distorted view in a small group. Take the study by psychologist Amy Cuddy.[35] Together with a colleague, she investigated whether your posture can make a difference mentally or physically. A powerful pose – with feet on the table or open arms – turned out to make a great deal of difference. The subjects not only reported feeling stronger in this posture, but the pose also had a biological effect; levels of the dominance hormone testosterone were higher, and those for the stress hormone cortisol lower. Cuddy's TED Talk about the topic was one of the most popular of all times and her book became a bestseller.

But if you take a look at the original study, you see that the conclusion has been drawn on the basis of a small group. Only forty-two people took part. When other researchers re-ran Cuddy's trial with two hundred people, the results were less spectacular. People felt more powerful, yes, but no difference in hormone levels was detected.[36]

It's not surprising that small studies are still conducted, especially in fields like neuroscience, because this kind of research tends to be extremely expensive.[37] But if we use these studies to gain an understanding of our psyche, health and development, we risk being wide of the mark.

The random sample, a solution to the problem?
After a five-day stay at the Institute for Sex Research, the three statisticians retired to write up their findings. During their discussions with Kinsey, they had chalked endless formulas and figures onto a blackboard to make him understand that his study was not representative. The professor had vehemently opposed

this view, but – unschooled as he was in statistics – he rarely had a good riposte.

Kinsey was nervous about the report the statisticians were about to write and decided to head for New York to ask George Gallup for advice. At the time, Gallup was *the* expert on opinion polls. In 1936, 1940 and 1944 he had predicted the winner of the American presidential elections correctly. But in 1948, he had put the wrong man forward. It had been the research by Gallup and other pollsters that had given the *Chicago Daily Tribune* the confidence to publish the bold headline announcing Dewey's victory.

The likely explanation for his own disgrace had become clear to Gallup in the meantime: quota samples. He had sent his pollsters into the country with a list of 'types', such as rural middle-class women. His interviewers had to collate a minimum number of questionnaires per type.

Gallup's method seemed a logical conclusion for the problems we saw earlier: no one was excluded in the sample and the quota saw to it that sufficient data was collected. The same idea is used by market research companies to this day. They often try to talk to people in every county or province and get a balanced picture in terms of gender and age. Once the figures have been collated, they are also corrected if certain groups have been over- or under-represented. When there have been too few women, for instance, the replies from female respondents will be weighted more heavily. A correction like this can help to make the data more representative.

And yet there is a persistent problem with Gallup's quota method. A pollster's report on how he went about his work demonstrates this clearly. In 1937, this data collector acquired his quota of lower-educated men by speaking to construction workers. He would join them during their lunch break. 'Do you approve or disapprove of a treaty with Germany?' he asked. 'How about you and you and you?'[38]

This 'method' did not work for people from the wealthier classes, he noted. 'You'd have to screw up your courage and go through a fancy part of town and try and figure out which house looked the most approachable.'

But what about the houses where guard dogs chased away the interviewer? Or the lower-educated men who were at home at lunchtime? They may have had a different opinion from their easily approachable counterparts, but they never found their way into this interviewer's dataset.

The fallacy in the quota method – and in the weighting methods of many present-day polling companies – is that there is an assumption that your opinion is influenced by just a few (easily measurable) factors, such as your income, gender and age. But alongside these factors you may also be affected by your personality, your dreams about the future, your youth, your sexual preference, your best friend ... Where does it end? So it is far from clear what influences your opinion and – here it comes – which factors a polling organisation should adjust for.

Hence, the quota sample would not have been a good alternative for Kinsey. But how *should* he have conducted his study? The three statisticians knew the answer: with a random sample. Kinsey might have been better off sticking a needle in a telephone directory, John Tukey argued, and interviewing all the people with a hole in their name. 'I would trade all your 18,000 case histories for 400 in a probability sample', he said.[39]

The random sample is still the holy grail of sample surveys. By giving everyone an equal chance to end up in the study, the hope is that you will achieve a good cross-section of the population.[40] An organisation such as a statistical bureau often has a file on citizens and can select a random group from this dataset. Following their humiliation in 1948, Gallup and fellow pollsters had started using random sampling. This was something Kinsey, who was now in a tight spot, wanted to learn about. Was a random sample really so much better?

Once in New York, Gallup coached the anxious Kinsey for hours on this method. Gallup assured him that, as far as the statisticians' criticism was concerned, things wouldn't be that bad. Because there was one big disadvantage attached to random sampling: not everyone would be available to take part in a survey.

4. Too few people want to take part

When Gallup and fellow pollsters tried to use random sampling it soon became clear that some people were not at home or did not want to take part. A random sample may have been scientifically justified, but pollsters like Gallup had only so much patience. Money had to be earned, so a somewhat less representative method would have to suffice.

Even if you approach a representative group, the problem of 'non-response' means that the group of people who end up taking part are not necessarily representative. With Kinsey's subject – sex – the likelihood that people would refuse to cooperate was particularly big. At university, for instance, boys would wait outside his door when he was interviewing a female student. Follow-up questions, they knew, were only asked when someone had sexual experience. So if she stayed longer than an hour, that meant – bingo – she was no longer a virgin.[41] No surprise that female students did not always feel like participating in Kinsey's study.

If too many people, then, say no, the random sample can be consigned to the dustbin at a stroke. Take this *New York Times* headline from 2015: '1 in 4 Women Experience Sex Assault on Campus.'[42] Twenty-five per cent of female students! A shocking result. But also, fortunately, likely to be too high. Let's take a look at the original research report, a 288-page volume.[43] It turns out that just twenty-seven colleges participated, only a small fraction of the total number in the United States. But what is more, of the 779,170 female students approached, only 150,072 decided to fill in the questionnaire. In other words, just 19.3 per cent actually participated.

All well and good; if the people who refused did not differ much from those who did in fact take part, there would be nothing to worry about. But there could be many reasons why they were different: women who have never been a victim of sexual assault might see less of a need to take the time to fill out the questionnaire. What, then, if the 80 per cent who hadn't participated were indeed those who had never experienced sexual assault or harassment? Then the percentage of victims would drop from 25 per cent to 5 per cent. If, on the other hand, you assumed that they would all have answered 'yes' – they had been assaulted – the figure would be as high as 85 per cent.[44] With topics as serious as sexual assualt, we need to take the numbers seriously. The researchers were very clear on these caveats; the *New York Times*, however, opted for a sensational headline.

This would be Kinsey's objection against the three statisticians who demanded he conduct a random sample: too few people would want to participate. And yet, not approaching potential nay-sayers is no solution either. Since, just as in the sexual assault poll, you want to work out what the effect of the refusal group would be. And this missing information not only made the Kinsey sex study unreliable, it also ensured that it was impossible to figure out *how* unreliable.

5. The margin of error is overlooked
Poor questions, exclusion, too small sample groups and non-response are four reasons why polls do not reflect reality as accurately as they

seem to. But even when the questions are more neutral than Switzerland and the sample is representative and large enough, we still face a problem that can never be solved: not everyone is being surveyed. Only a section of the entire group is being interviewed; that is the whole idea of a sample. That smaller group will rarely look *exactly* like the whole population. If Kinsey had used a random sample, he would have had a slightly higher number of homosexuals at one time than at another. Or fewer adulterers. Simply because chance dictates who ends up in the group.

For that reason a poll always has a margin of error. That bandwidth indicates how much reality can deviate from the result.[45] The larger the sample – is the rule of thumb – the smaller the margin. The exact size of the margin can be calculated with a formula, but easier still: you can look up an online calculator on a website such as goodcalculators.com, which works out margins for random samples.

Suppose Kinsey *had* selected his sample randomly. At the point that he concluded that 50 per cent of his respondents had been unfaithful, how big would his margin of error have been? If he had only spoken with 100 men, the percentage could have ended up 10 percentage points higher or lower.[46] A bandwidth of a hefty 20 percentage points. But because he had some 5,300 men in his sample, the margin of error would have been just 1.3 percentage points.

Margins of error in samples are often overlooked in the media, most notably when they are about elections. Election polls can be wrong by a couple of percentage points, but a small swing is sometimes given huge significance in newspaper columns and on talk-show sofas.

And whereas, in 2016, many newspapers argued that the polls for the American elections had been mightily wrong, they weren't as far wrong as they seemed if you take into account the error margins. In some states the pollsters had certainly messed up. In the state of

Wisconsin, Trump did 6 percentage points better than the polls of the Marquette Law School had predicted; in the suburbs of Milwaukee, as much as 10 percentage points better.[47]

But by and large the poll predictions had been fairly accurate. In the end, in the popular vote – the vote among the entire American population[48] – Trump scored only between 1 and 2 percentage points higher than the polls predicted, within the margin of a renowned pollster such as ABC News/*Washington Post*, who reported a margin of 4 percentage points.[49] So there was nothing surprising about Trump's victory if you had taken into account the error margins. What's more, the difference between the polls and the result was even less than when Obama won in 2012, when no one complained about the figures.[50] It had not been the pollsters who had got it wrong in 2016, it had been the media.

The lesson? When collecting numbers it is more or less universally the case that the outcomes cannot be totally precise. Do not view them as an exact representation of the truth, but rather as if you are looking through frosted glass: you can see the contours, but what's within will never be sharply in focus.

Intermezzo: When anchor woman Dionne Stax talks about percentages

'A quick comment,' said Dionne Stax on Dutch TV on 18 March 2015.[51] 'I should really say "percentage point" when I want to be totally correct, but that's not what we're going to do tonight. I'll just stick to per cent. So you know.'

You can set your watch by it: during every election night, people complain about the incorrect use of the word 'per cent'. It was no different during the Dutch regional elections. Stax discussed the results on TV and was soon at the receiving end of a particular Twitter criticism. The reason: she mixed up 'per cent' and 'percentage point'.

What is the difference between the two? Let's say one party got 5 per cent of the votes in previous elections and is now up to 10 per cent. An increase of 5 per cent, Stax would have said in such a case. But that is in fact wrong: the proportion has doubled, so increased by *100 per cent*. If you want to say something à la Stax, you should call it an increase of 5 *percentage points*.

6. A particular outcome matters to the researcher

In 1954, four years after their visit to Kinsey's institute, statisticians Mosteller, Cochran and Tukey published their 338-page critical report about the sex study. Kinsey had carried out impressive work, they concluded, but the sample was not a fair reflection of American men. Kinsey had meanwhile published a study about the sex life of women using the same method. Once again the sample was not representative, so once again it gave a distorted view. But it made no difference. 'Most Americans could hardly have cared less what academics thought. They wanted to hear what Kinsey had found out about American women', Kinsey's biographer James Jones wrote in 1997.[52]

Even today, Kinsey's sex study encourages fierce discussions. These tend not to be about the representativeness of the study, but about four striking tables in Chapter 5 of Kinsey's report on men. They deal with 317 boys – the oldest is fifteen, the youngest a mere two months old. The first table shows which percentage had experienced an orgasm; the second the length of time it took to reach an orgasm (3.02 minutes on average); the third and fourth tables include boys who have had several orgasms during the observation period, which might last as long as twenty-four hours. The accompanying text states that this data was collected by nine different men. But in 2005 this turned out to be a lie: there had only been one source who had supplied this data.[53] Kinsey had wanted to protect this man by pretending that there had been multiple sources.

What was the story? As a small child, this Mr X had had sex with his grandmother and his father.[54] It was the beginning of a sex-obsessed life. Kinsey's colleague first wrote about this man in 1972, who, by the time they came into contact with him, 'had had homosexual relations with 600 pre-adolescent males, heterosexual relations with 200 pre-adolescent females, intercourse with countless adults of both sexes, with animals of many species [. . .]'.[55] Mr X had kept detailed records of his various encounters.

Kinsey saw the records as a scientific goldmine. 'I congratulate you on the research spirit which has led you to collect data over these many years', he wrote. Mr X, a civil servant whose job required him to travel a great deal, had drilled holes in the walls of hotel rooms to spy on his neighbours, recording all sexual activity he encountered. '[I am] very much interested in your account of hotel observations', Kinsey wrote. He saw no problem in the use of the data. As a researcher, he believed it was his duty to collate facts. Not to pass moral judgement.

Kinsey missed the point; as a researcher you *always* pass moral judgement. Researchers choose which topic is important, how they deal with respondents, what they ultimately do with the information they have gathered. Kinsey's lie that the data came from several men was a scientific mistake; accepting figures about child abuse was, in the eyes of many, a moral one. By treating Mr X as a colleague, Kinsey implicitly approved of his behaviour.

This was not the only snag. Kinsey had a mission. Behind the scenes, this ostensibly objective professor with the bow tie had been wrestling with his own sexual identity for decades. According to James Jones's biography, Kinsey had affairs with men, experimented with S&M and encouraged his university colleagues to have open marriages. The conservative sexual norms of the time prevented people from being themselves, Kinsey believed. He even wondered if paedophilia was as bad as people thought. In some cases, Kinsey had told a colleague, sexual contact between an adult and a child could even be beneficial.

When the film *Kinsey*, starring Liam Neeson, was released in cinemas in 2004, the discussion about Kinsey's 1948 sex study flared up again. Advocates of sexual freedom called Kinsey a trailblazer for the sexual revolution, the pill, abortion and gay rights. Opponents blamed him for having made abhorrent sexual norms acceptable. Whichever side you are on, there's no getting around the fact that Kinsey's data was not objective. It was influenced by a mission to break open sexual norms. Never just ask *how* the figures were collated, therefore, but also *who* collated them.

In the case of Kinsey, his unrepresentative figures confirmed what his gut feelings told him: that the actual behaviour of people was quite different from what the norms prescribed. His research was activism, wrapped in a scientific cloak of graphs and charts.

CHAPTER 4

SMOKING CAUSES LUNG CANCER (BUT STORKS DO NOT DELIVER BABIES)

It was 1953 and the tobacco industry was in trouble.[1] Shares in Philip Morris & Co., United States Tobacco Company and other manufacturers suddenly plunged in value. The immediate cause was a publication by cancer researcher Ernest Wynder and colleagues, who had painted tar from cigarettes onto the shaved backs of white mice using a camel-hair brush.[2]

The results of this trial were shocking: 44 per cent of the mice in the test group developed cancer. Out of the eighty-one mice that had been painted with tar, only 10 per cent were still alive after twenty months. Not a single incidence of cancer was found in the non-tarred control group, and 53 per cent were still alive after twenty months. The *New York Times*, *Life* and the tremendously popular *Reader's Digest* had written with disquiet about the trial. The latter under the grim title 'Cancer by the Carton'.

The tobacco magnates were no longer able to ignore the furore and met up under the high ceilings of The Oak Room on New York's Central Park in December of that year.[3] In this renowned restaurant, they sought to forge a plan to protect the industry against critical researchers. And who better to help them than the very man joining them at the table: John Hill. Hill was the CEO of Hill and Knowlton, one of America's most powerful PR firms. With his help, the tobacco magnates wanted to convince the public that there was no scientific foundation for the accusations coming from Wynder and his colleagues. They would demonstrate that all these worries about cigarettes were ridiculous

So on 4 January 1954 the big tobacco manufacturers launched the Tobacco Industry Research Committee.[4] In full-page advertisements in more than four hundred different newspapers they assured the public that their products were not harmful.[5] Throughout the hundreds of years that people had enjoyed tobacco, they argued, critics had blamed it 'for practically every disease of the human body'. Time and again the criticisms would be dropped for lack of medical evidence, the committee stated. But the fact that there was even a suspicion of harm now caused deep concern to the manufacturers, they wrote. With their joint industry committee, they would contribute to research into 'all phases of tobacco use and health'.

It was the beginning of a conspiracy that would last almost fifty years and would cost countless lives. The American Department of Justice would later argue that, on that infamous December day, the magnates had decided 'to deceive the American public about the health effects of smoking'.[6]

But the tobacco industry was not alone in the deception. Thousands of scientists were complicit in the deceit.

Lying with statistics

During the same year that the tobacco industry's full-page advertisements appeared, Darrell Huff published *How to Lie with Statistics*.[7] This 142-page book would become one of the most popular titles about numbers ever. Huff was no statistician, but a journalist with an irrepressible curiosity.[8]

His earlier books had been about photography, careers and dogs, and now he had got his teeth into the misuse of numbers. 'The crooks already know these tricks; honest men must learn them in self-defence.' The book became a roaring success and more than 1.5 million copies of the English version alone were sold.

It's my favourite book about numbers. Sprinkled with humour, Huff writes about mistakes that are still made today, such as non-representative polls and misleading graphics. He also writes about another classic mistake: confusing correlation with causality. The mistaken idea that, because there is a link between two things, one automatically causes the other.

For instance, Huff deftly shows that you can make a good estimate of the number of babies in a house by counting the number of stork nests on its roof. In other words, there is a link between babies and storks. But, spoiler alert, children are not delivered by black and white birds. The link between the two (correlation) does not mean that one causes the other (causality). It is highly likely there is another factor that influences both issues. 'Big houses attract big, and potentially big, families,' Huff writes, 'and big houses have more chimney pots on which storks may nest.'

Recognising this mistake is not just important for statisticians, but for all of us. Many important decisions are based on a presumed causal relationship. The government chooses to introduce cuts because it thinks cuts lead to a smaller public debt. A smoker stops smoking because doctors allege that he will get lung cancer if he doesn't. And I try to fly as little as I can because experts tell me this is better for the environment. The idea is that if you know how something has been caused, you can also change it.

But you should not confuse correlation with causality. We saw this mistake crop up earlier, when politicians claimed that someone's skin colour determined his or her IQ score. And when psychologist

Amy Cuddy argued that particular postures have an effect on your hormone levels.

But nowhere is the causality error more ubiquitous than in news about health. Drinking gin and tonic will reduce your hay fever symptoms,[9] you are more likely to get an STD if you shave your pubic hair,[10] and dark chocolate is good for your heart[11] – these are just a few of the reports that swamp us daily. Such statements tend to be exaggerated. This is not just due to the media, which likes to circulate overblown reports; in fact, the problem often starts in the press departments of universities, which disseminate health studies. Researcher Petroc Sumner and colleagues looked at press releases from 2011 on biomedical and health-related science, published by twenty universities in the UK. They found that roughly 33 per cent of the releases exaggerated causal claims[12] and that around 80 per cent of the news stories adopted such exaggerations.

When, as a news consumer, you can no longer have total confidence in journalists and scientists, how can you separate facts from fiction? How do you know whether smoking causes lung cancer, for instance? *How to Lie with Statistics* gives us something to go on. In it, Huff describes three types of, what I like to call, 'cocky correlations' – correlations that pretend to be something more than they are: causal relationships.

1. It's a coincidence

A recipe book. This was the source doctors Jonathan Schoenfeld and John Ioannidis used for their cancer studies analysis.[13] They chose some random recipes from *The Boston Cooking School Cook Book* and noted down the first fifty ingredients they came across. With this list, they plunged into PubMed, the medical research archive. Their first finding was rather curious: forty out of the fifty ingredients turned out to be related to cancer in one or several studies. 'Is everything we eat associated with cancer?' the researchers wondered.

Their next finding was plain bizarre. Often, for the same ingredient, an increased as well as a decreased cancer risk was found. If one study found that wine was good for you, for instance, another study could be found that claimed you'd be better off not touching that glass in front of you.

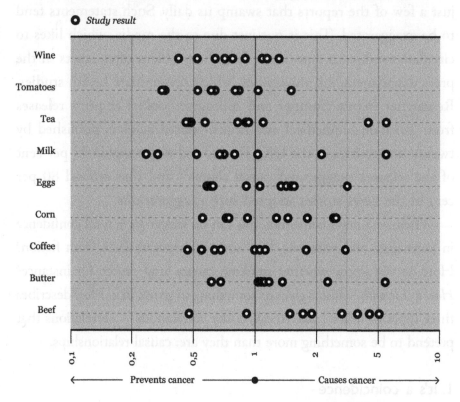

Source: Schoenfeld and Ioannidis (2013)

Schoenfeld and Ioannidis decided to confine their study to the twenty ingredients for which there were at least ten studies available. Out of those twenty ingredients they found contradictions in the conclusions for seventeen products, from tomatoes to tea, from coffee to beef.

The results could not all be correct, but how did the researchers of these studies come to their conclusions? Huff's first kind of cocky correlation gives a possible explanation: *it was a coincidence.* The story of an octopod fortune-teller shows what happens when coincidence and correlation coincide.[14] In 2010, Paul the Octopus predicted the results of eight World Cup matches. Time and again he used his tentacles to open the correct food box, the box with the flag of the football team that would win the subsequent match. And time and again hordes of journalists waited excitedly for his prediction. When the Netherlands ended up playing the final against Spain, Paul anticipated the Dutch defeat. The octopus became a celebrity: he was made an honorary citizen of the Spanish town of O Carballiño, became ambassador for England's attempts to stage the World Cup in 2018, and was seen as a 'symbol of Western decadency and decay' by the President of Iran, Mahmoud Ahmadinejad.

But what if Paul had simply had a lucky break? The odds of him predicting eight matches correctly by pure chance is equal to the odds of always getting heads when you throw a dice eight times: one in 256, or 0.4 per cent. A small chance, but the odds you might win the Lotto are almost 200,000 times as small – one in 45 million.[15]

It becomes even less spectacular when you know what other animals were in contention for the role of World Cup soothsayer. What to make of Leon the Porcupine, Petty the Pygmy Hippopotamus and Anton the Tamarin? They too made predictions on the World Cup games, but were less lucky than their colleague Paul. If you allow enough animals to predict outcomes, there will always be one that gets it right.

It's the same with correlations. If you look long enough, you will always hit upon a relationship. No one illustrated this better than analyst Tyler Vigen. He became famous as a result of the strange correlations he publicised on his website Spurious Correlations.[16] He discovered, for example, that the increase in the yearly number of people who drowned as a result of falling into a swimming-pool corresponded almost exactly to the number of films that feature Nicolas Cage. And the trend in cheese consumption looked frighteningly close to the number of people who die by becoming entangled in their bedsheets.

Vigen's correlations are patently nonsense, which makes them so comical. What's not nearly as funny is that correlations in health research may have arisen through chance just as easily.

Cartoonist Randall Munroe showed how this worked in his webcomic xkcd.[17] A stick figure with a ponytail comes running into the cartoon, exclaiming: 'Jelly beans cause acne!' In the next frame, two scientists – one stick figure with lab glasses and one clutching a piece of paper – present the results of their research: there is no link. 'I hear it's only a certain colour that causes it,' the ponytail responds. The scientists come back again, this time with the message that there is no link with purple jelly beans. Nor with brown, pink, blue, teal, salmon, red, turquoise, magenta, yellow, grey, tan, cyan, mauve, beige, lilac, black, peach or orange jelly beans. But they did find a link with one colour. The last frame shows the front page of a newspaper: 'Green jelly beans linked to acne!'

Earlier we saw the problem of samples that are too small; this cartoon shows two additional, prevalent problems in science. The first is publication bias. You tend only to hear about studies that have found a significant correlation. In many fields of study the mantra is: not significant, not important. This applies not only when you want to make sure your research is picked up by the media, but also when you want to publish in a scientific journal. Many studies with a null result consequently stay in the drawer, giving scientific literature a distorted image. Because researchers want to publish their work, they will look for clear correlations in data. This does not sound wrong in itself, but as with the jelly bean cartoon, if you look long enough, you will always find something.

The newspaper front page in the cartoon also says: 'Only 5 per cent of a coincidence!' By this, cartoonist Munroe is referring to the so-called p-value, which measures to what extent the outcome came about as a result of coincidence. The renowned statistician Ronald Fisher was responsible for ensuring that, during the twentieth century, the p-value became *the* method to measure the significance of a correlation.

Say you want to investigate whether there is a causal link between green jelly beans and acne. You can find this out by conducting a trial, as Archie Cochrane did: you divide your test subjects into two groups. One group is given a green jelly bean daily for a month, the other a green sugar pill. Out of the group that received the placebo, 10 per cent suffer at the end of the trial. In the jelly bean group more people have acne, but this could of course be sheer coincidence.

Obviously, if in this group, 100 per cent of the test subjects have spots, it is very unlikely that this happened by chance. But is 90 per cent high enough? Or 50 per cent? You have to draw the line somewhere. The p-value is the likelihood that, in case the jelly beans do not actually cause acne at all, you will still end up finding a certain higher percentage of acne patients in the jelly bean group. If this likelihood is below the agreed threshold – often 5 per cent – then the likelihood that you will detect this percentage of patients is so small that you can call the correlation 'statistically significant'.

But it can still mean that jelly beans do not cause acne at all. With a p-value of 5 per cent you will still find a surprising result within 5 per cent of the studies. The odds that you will win the lottery are much smaller, but here too there are winners.

And now we get to the second number problem in science. For a long time, in many social sciences, there has been a monomaniacal focus on p-values. Scientific journals have preferred to publish significant results only; besides, many researchers have to live by the motto 'publish or perish'. If you don't publish enough, you're in the doghouse. Which is why some scientists frantically began to dig for p-values that were as low as possible. This is called p-hacking.

Former Cornell University professor Brian Wansink took p-hacking to a new level. He became famous through studies that, according to him, showed that children are more likely to choose apples if you stick a Sesame Street sticker on them[18] and that people eat less from a small plate.[19] His findings received much attention in the media, including in the *New York Times*, and he led a nutrition centre at the Ministry of Agriculture under President George W. Bush.

But his work turned out to be full of holes. Leaked emails in 2017 showed in no uncertain terms how Wansink and his colleagues had set to work. One of his researchers, for instance, emailed that she had analysed data from an all-you-can-eat-restaurant, but this had produced no results. Wansink's email reply: 'I don't think I've ever done an interesting study where the data "came out" the first time I looked at it.'[20] He had an idea for his colleague. 'Think of all the different ways you can cut the data and analyse subsets of it to see when this relationship holds.' In other words, study all the jelly beans until you find a colour that is linked to acne.

Suddenly it does not seem so strange that Schoenfeld and Ioannidis discovered that so much of our food had a link with cancer. Thanks to publication bias, the studies that did not find a relationship never saw the light of day, and researchers were able to p-hack for as long as they came across a correlation with a low enough p-value. Whether this correlation was positive on one occasion and negative on the next no longer made much difference. As long as it was significant.

2. A factor is missing

Once Archie Cochrane had received the shipment of yeast from the Germans in August 1941, the number of patients with oedema in the prison camp soon plummeted. And yet we cannot say for sure whether it was the yeast that was behind the sudden drop in cases, for when Cochrane submitted his request to the Germans, he had begged not only for 'A lot of yeast at once', but also for 'an increased diet as soon as possible'.[21] Both requests found a willing ear. The yeast arrived, and within a few days, the prisoners were also given more to eat, and they now consumed a – still paltry – amount of eight hundred calories a day. The cause of the sudden drop in oedema cases? It could well have been the richer diet.

There was another issue. As described earlier, Cochrane called this his most successful *and* his worst trial, because the groups had been too small. He gave yet another reason: he had tested the wrong hypothesis. Cochrane had made the assumption that beriberi had been the cause of the swollen ankles and knees. That's why he had experimented with vitamin B (yeast). But in his autobiography he writes that hunger oedema was the most probable cause, not beriberi. In the case of hunger oedema, the answer is not an increase in vitamin B, but more food. Why did the patients in his yeast trial recover? That's a 'mystery', writes Cochrane, but he suspected it was as a result of the protein in the yeast.

This takes us to the second cocky correlation: *a factor is overlooked that affects both 'cause' and 'effect'*. This is exactly what we see happening in Cochrane's story. Because of the yeast, the prisoners consumed more vitamin B ('cause') *and* suffered less from oedema ('effect'), but that did not mean that the vitamin B was the cure of the oedema. It's comparable to Huff's example of the storks and the babies. This time it was not the size of the roof, but the extra food that was the third factor.

Let's look at another example. In his book Huff describes a study into smoking and school exam results. Smokers, the study revealed, achieve lower grades. Should students stop smoking? Nonsense, Huff thought. Here too, other factors could be at play that had an effect on someone achieving lower grades *and* on someone smoking. Perhaps more sociable people tended to smoke, and because of their social lives they would not spend as much time with their noses in books. Or was the difference to do with being extroverted or introverted? 'The point is that when there are many reasonable explanations, you are hardly entitled to pick one that suits your taste and insist on it.'

The same mistake was made in 2015 in a large Dutch study of more than 37,000 breast cancer patients.[22] According to the press release, the researchers concluded that women who had had a lumpectomy, often combined with radiotherapy, tended to live longer than patients who had undergone a mastectomy.[23] The claim received much media attention and in no time the Dutch Breast Cancer Association was swamped by questions from worried women. Had their mastectomy been a mistake? Should they be getting radiotherapy after all? Messages to reassure people soon appeared on hospital websites,[24] and the authors of the study would later emphasise that they had not found a causal link.[25]

The issue was that many other factors played a part, factors that were linked to why a particular treatment had been chosen ('cause') and the survival rate ('effect'). For example, if a patient had another illness – such as heart failure – then a mastectomy was a more common option,[26] the idea being that radiotherapy would be too major an intervention in an already weak physique. That this group tended to die sooner might have had nothing to do with the surgery, but was connected instead with their worse general health.

3. There is a reverse causal relationship

The third and last cocky correlation Huff discusses is when *the causal link works the other way around*. When it's raining you see many people with umbrellas. In this instance, can we say that umbrellas have caused the rain? Of course not. It's the rain that has led to all those umbrellas.

But cause and effect are not always all that clear, Huff shows. When a wealthy person owns many shares, have they become rich as a result of those shares? Or were they able to buy them because they already had plenty of money? Both can be true. The causality can work both ways – someone is rich, buys shares, becomes richer, buys more shares, etcetera.

The same goes for the 'obesity paradox', the finding that overweight people sometimes have better survival rates than people with a 'normal' weight. This is surprising, because you tend to hear that obesity is bad for your health. Researchers concluded that obesity must have a protective function that keeps you alive for longer.

But an important factor was overlooked: when you are ill, you lose weight. The lower weight was not necessarily the cause of bad health, but the consequence of it. This conclusion was confirmed in a study from 2015, in which weight loss was adjusted for.[27]

So remember that correlation does not automatically imply a causal link, because coincidence (cocky correlation 1), a missing factor (cocky correlation 2) or reverse causality (cocky correlation 3) may have played a part.

But how do you know when there *is* a causal link? More specifically, how did we ever find out that smoking leads to lung cancer?[28]

Intermezzo: When everyone suddenly gets hot under the collar over bacon

In the spring of 2015, news about processed meat products such as sausages and bacon made the headlines.[29] NOS, the Dutch news organisation, reported: 'People who eat processed meat on a daily basis are almost twenty times as likely to develop bowel cancer.' Internationally, many other media outlets also talked about this news. Or, as comedian Arjen Lubach put it: 'Everyone took part in the game of how can we report this as carcinogenic as possible?'[30] Take the headline in the Dutch edition of *Metro*: 'Bacon is as cancer-inducing as smoking.' Going at it hammer and tongs the next day with 'Can I still eat without dying?' (If you manage to do this, you'd be the first, Lubach remarked.)

NOS had overegged the message a bit, too: that 'almost twenty times' should have been 'almost 20 per cent'. And yet even the media that produced the right figures also participated in the scaremongering. Fair enough, because an increase of 20 per cent seems considerable.

But an important detail was missing in much of the reporting: 20 per cent *of what?* If you look at the data, six in a hundred Dutch people will get bowel cancer at some point in their life.[31] According to the World Health Organisation, this percentage drops by 18 per cent – this is where that 'almost 20 per cent' came from – if you stop eating processed meat.[32] From six to five in a hundred.

You often see such reporting around health news; you read about the *relative* risk (almost 20 per cent), but nothing about what this means in *absolute* terms (one in a hundred).

How Hitler could have saved the lives of millions of smokers

How did the research into smoking and lung cancer begin? The experiment conducted by Wynder and his colleagues, who spread tar onto the backs of mice, had rattled the tobacco manufacturers in 1953. But scientific research into the health risks of smoking was much older. As far back as 1898, the German medical student Hermann Rottmann wrote about a possible link between smoking and lung cancer, and in 1930 the German physician Fritz Lickint was one of the first people to publish statistical proof of the correlation.[33] An Argentinian doctor, Angel Roffo, did the first experiment on animals at around the same time, whereby he applied tar onto the ears of rabbits. A sickening drawing shows how a velvety brown ear is dotted with raspberry-pink growths. Roffo published hundreds of articles about smoking and lung cancer, primarily in German academic journals.

It's no coincidence that the early research into the effects of smoking was strongly linked to Germany. During the 1930s, it was the most developed country in the field of medicine. What's more, there would be no leader more vehemently opposed to smoking than Adolf Hitler. He even claimed that national socialism could never have triumphed if he himself had not stopped smoking in 1919. It was the Führer, not the cigarette, that needed to be in control of people's bodies. And so this threat, just like the Jews, had to be kept at bay.

In 1939, the German researcher Lickint published *Tabak und Organismus*, a 1,200-page book in which he summarised more than 7,000 studies on the effects of tobacco. This and other meta-studies (research into research) led to a consensus among experts. At the beginning of the 1940s most German doctors and officials agreed: smoking was dangerous.

But it was not this German study that helped us to understand that smoking causes lung cancer. When Wynder and his colleagues published the results of their mice experiment, they were greeted as pioneers. Similarly, research by the British epidemiologists Richard Doll and A. Bradford Hill from 1952 was seen as revolutionary.[34] Even today, these Anglo-Saxon scientists are considered the founders of research into smoking. And yet, although their research may have been more advanced, the German scientists were ahead of them by at least ten years.

But after the war, the German studies disappeared from the scientific consciousness. Many scientists from Germany had not survived the war. More importantly still, medical research by Germans had left a bad taste in the mouth.

What does this clarify? That scientific progress does not always happen in a straight line. Progress is made, only to find itself back at the starting point again after a few years. An ironic aside: one of history's biggest mass murderers could have saved the lives of millions of smokers with his anti-smoking propaganda.

But the unsavoury image of German research is not the only reason that the association between smoking and lung cancer remained hidden for so long.

The most insidious marketing trick

It was 1970, and at a high school in Kansas City all the pupils were summoned to the assembly hall to listen to a young man in a striped shirt and white shoes. He was there on behalf of the tobacco industry with a simple message: smoking is not for children. It was something for grown-ups, as was sex, alcohol and driving. Something teenagers should not even think about.

It seemed a well-intentioned message, but if there was anything the kids were thinking of now it was cigarettes. And if there's one thing teenagers fall for, it's something that's forbidden. Something that's only for adults.

Years later, one of the pupils in the hall, Robert Proctor, wrote about this assembly in his book *Golden Holocaust*.[35] The young man, he said, was part of an insidious campaign to encourage children to smoke.

Proctor had become a historian by then and had devoured millions of classified documents from the tobacco industry. He found an array of dubious practices. It turned out to be a deliberate decision to target children. These 'pre-smokers', 'the cigarette business of tomorrow' or the 'replacement smokers', were substitutes for the smokers who had been forced to stop smoking (read: they had died). In 2000, Philip Morris International sent 13 million book covers to American Schools. Pupils were given the opportunity to cover their books with an image of a cool snowboarder and the text 'Think. Don't smoke.' Tobacco brands approached pupils not only via schools, but also through parents. Information leaflets exhorted parents to talk about the dangers of smoking with their children.

The tobacco industry is known for using smooth slogans ('I'd Walk a Mile for a Camel'), strong role models (the Marlboro Man), and for being the first to deploy billboards, product placement in Hollywood films and impulse buys in supermarkets. But inconspicuous, cunning marketing tricks were what truly separated the tobacco industry from other companies. Proctor unearthed in secret memos and other documents how cigarettes were made increasingly addictive over the years, by adding liquorice for instance, to make the smoke taste sweeter, or ammonia, which made the nicotine more addictive.[36]

And there was one marketing trick that was the most villainous of all. It was a ruse cooked up in The Oak Room in 1953 and, since then, it has misled millions of people about the effects of cigarettes. The scheme was best summarised by John W. Burgard, marketing director for one of the big tobacco brands, who – obviously in a classified document – wrote: 'Doubt is our product.'

The tobacco magnates' objective was not to demonstrate that smoking is good for you. The existence of doubt about the effects of tobacco was already enough. Since the meeting at The Oak Room, the Tobacco Industry Research Committee – later the Council for

Tobacco Research – would do everything in its power to sow confusion about the findings of scientific research into smoking. The club was only disbanded in 1998, after a legal agreement between the tobacco industry and the public prosecutors of forty-seven American States. In the decades before, the tobacco industry had spent hundreds of millions of dollars on medical research.

The commission's research grants appeared to go towards funding studies about 'tobacco and health', but in reality were seldom dedicated to this. 'The goal was really to look in such a way as not to find,' Proctor wrote, 'and then to claim that despite the many millions spent on "smoking and health" no proof of harms had ever been uncovered.' He came across hundreds of press releases with the scientific mantra 'more research is needed'. Or, as one of the tobacco magnates put it, 'research must go on and on'.

Not only could the tobacco industry now suggest that it took science extremely seriously indeed, but in addition its funding of grants to researchers from reputable universities such as Stanford and Harvard helped its image. At the same time, it set up 'a stable of experts', with scientists who would write 'industry-friendly' articles, or testify in court, if necessary.

And so we come back to Darrell Huff. He may not have been a scientist, but the author of *How to Lie with Statistics* was the perfect fit for the stable. Who could speak more juicily about numbers than Mr How-to-Lie-with-Statistics himself?

And so, on 22 March 1965, he testified to the American Congress in a hearing about cigarette advertising and packaging. The last thing you should do, he said, was to confuse the correlation between smoking and bad health with causality.

Intermezzo: The graph that makes sure you never age
Florence Nightingale knew how to convince the government with

graphics. But they can also be used to sow doubt. In 1979, the Tobacco Institute, an institute financed by the tobacco industry, published a graph showing the development of different types of cancer. Scientific studies suggested that both the number of smokers and the number of cancer patients had gone up over the years.

The graph was supposed to show that this was not necessarily the case. It gave a picture of the proportion of patients with mouth and throat cancer, bladder cancer and cancer of the oesophagus. The result looked so messy that it was hard to argue there was a consistent increase. But what was missing in the graph? Sure enough, the most important effect of smoking: lung cancer.

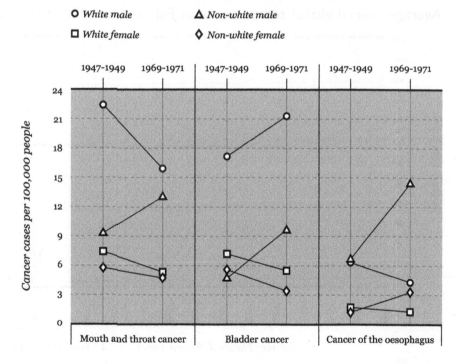

This graph was circulated by the Tobacco Institute in 1979.
Source: Proctor (2011), figure 29.

It's not only the tobacco industry that sows doubt with graphics. On 14 December 2015 the *National Review*, a conservative American magazine, tweeted: 'The only #climatechange chart you need to see.'[37] The image showed the temperature since 1880. The outcome? The average temperature had barely changed over the past 135 years. The line showing temperature change was as flat as the heart monitor output of a patient who's just died.

My instinctive reaction was that the data must be flawed, because countless readings show that temperatures are rising.[38] The *National Review* must have cooked up those figures; there was no other explanation. But no, the data was correct. It came from a reliable source: NASA, the American space agency.[39]

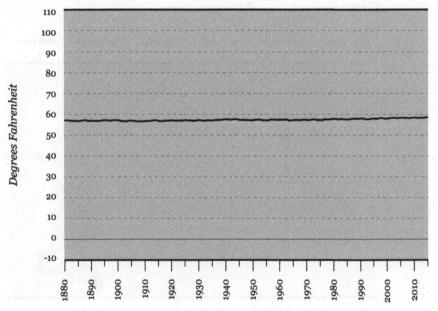

Average annual global temperature in Fahrenheit 1880–2015

Source: *National Review* tweet on 14 December 2015

Let's have another look. The graph features an unambiguous title and includes labels for both the axes; it meets all the requirements for graphs you learnt at school. The period on the horizontal axis, 1880 to after 2010, seems perfectly good for conveying long-term change. And there seems nothing wrong either with the scale of the vertical axis: -10 to 110° Fahrenheit, which converts to -23 to 43° Celsius. No absurd temperatures; there are places on earth that can be that cold (Siberia) or hot (Las Vegas).

And yet something is awry on the vertical axis. It's not dealing with the temperature at one place at one time. It's giving the *average* temperature for the whole world. And in this, a few *tenths* of one degree already makes a difference. Climate experts agree that an average warming of less than 2 degrees Celsius may have catastrophic consequences.[40] It's impossible to detect such a change in this graph, because the scale of the vertical axis is so small.

It's as if I look at the graph below and conclude that I have not aged a single day during the past thirty-one years.

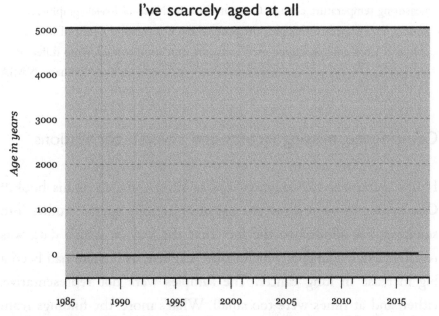

If you alter the axis in the climate graph, you suddenly end up with an entirely different picture.

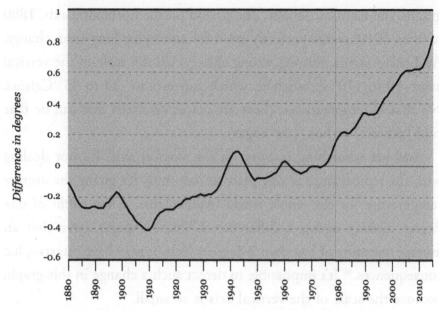

This graph shows the annual difference between the average temperature in degrees Celsius and the average for the period 1951–80.[41] This method of measurement is also called anomaly and is the standard in climate science for measuring temperature change. Compared to the *National Review* graphic, more aspects have therefore changed: the scale of the y-axis and the unit of measurement. If I had only amended the y-axis, the conclusion would be no different.

Source: NASA

Coincidence, missing factors and reverse correlations

Huff's statement to Congress was as silver-tongued as his book.[42] One by one he went down the list of objections against studies into smoking. He alluded to the fact that the way in which data was registered had changed, which made it appear as if there had been a big increase in lung cancer. The samples were not representative, either, and at times were too small. What's more, the findings from animal tests should not have been applied to people without further research. He must have been thinking of the authoritative study by

Wynder and his team, wherein they painted tar on the backs of mice, when he argued: 'Mice aren't men.'

He constructed his plea in such a way as to build up to his chief objection: 'If, in spite of these differences, we accept some kind of association between smoking and health, we fetch up against a final and crucial question.' Does the correlation between smoking and cancer automatically mean that there is a causal link? No, Huff argued, launching into a discussion about storks and babies.

He recited the three mistakes concerning correlations that he'd written about in his book. Earlier in his statement he had already argued that the differences in cancer cases between smokers and non-smokers might have been 'statistically significant', but could just as easily have been due to chance. He also seemed to suggest that the causal relationship could run in reverse, when he said: 'If Yale graduates really do have more money than most of us, is it because they went to Yale? Or is it that Yale generally attracts men from wealthy families?'

Huff was not the first to point to the possibility of a reverse causality. Ronald Fisher, the statistician who had popularised the p-value, had already suggested the same. 'Is it possible, then, that lung cancer [. . .] is one of the causes of smoking cigarettes?' he wrote in a pamphlet in 1959.[43] Even before the disease has been detected, Fisher suggested, people already have minor inflammations. Just as people light up when things get tough – a delayed train, a frustrating meeting – people may smoke because their lungs are giving them trouble. 'And to take the poor chap's cigarettes away from him would be rather like taking away his white stick from a blind man.'

But Fisher eventually found another, more probable explanation: a factor was missing. He was convinced that genes could explain almost all the differences between people. If you had particular genes, Fisher thought, you would be more likely to take up smoking.

Darrell Huff did not talk about genes in Congress, but he too thought smokers had a different make-up to non-smokers. They were more often overweight and drank more beer, whisky and coffee. In addition, more of them also tended to be married, be treated at the hospital more often and change jobs more frequently. You could not just select one of these explanations and ignore the rest.

When do you know enough?

Is there such a thing as truth? What remains of numbers after all the nuanced qualifications about standardising, the mistakes in data collection, and the misleading and wrong analyses we've seen here? Would it be better to disregard numbers and shroud ourselves in blue smoke as if we were the advertising wizards in *Mad Men*, because we cannot know what smoking does to us anyway?

Huff and Fisher's arguments are founded on three kinds of cocky correlations. There was a correlation, but this correlation was not necessarily causal. If the physical health of women who had had a mastectomy differed from women who had not, why did this reasoning not apply to smokers and non-smokers? How can we be sure that the research into smoking and lung cancer doesn't suffer from publication bias, with the null results studies remaining in the drawer? And was there some truth in Fisher's reverse causality, which had also explained the obesity paradox?

This was the tobacco industry's clever stratagem: they came up with arguments that were entirely valid in other contexts. It is of course feasible that the results of a particular study came about through chance. Even if this had not been the case, there might still have been other factors that were not included. Fisher argued in his pamphlet that there was only one way to exclude these alternative explanations: an experiment. But he knew that the medics and the general public considered it unethical to allow people to smoke if it might do them harm. So the trials were not conducted on people, but on animals. And that's where Huff's argument came in: 'Mice aren't men.'

Huff and Fisher thus spun a web from which there was no escape. With these arguments it was simply impossible ever to come to a sound conclusion. And that is exactly where the tobacco industry wanted the discussion to be: in an endless tunnel, where you could keep on clamouring for more research and where you could never draw any conclusions.

This is the big challenge that faces science: trashing a causal link is easy, proving its existence is extremely difficult. How *do* we know that smoking causes lung cancer? Huff and Fisher's arguments held water, but only if you looked at individual studies. One study, however well conducted, is never enough to prove something. A particular group in a particular country has been monitored at a particular time and you can still say that the outcome has been the result of chance. That's why it's so problematic when newspapers write that something has been 'scientifically proven' on account of just a single new study. And it's an equally bad idea to rely on one poll to predict the outcome of an election.

Science is not about loose studies, but about the *accumulation* of studies. And by the time Huff was questioned in Congress in 1965, that accumulation was enormous. The magisterial overview study *Tabak und Organismus* from 1939 may have been forgotten, but the burden of proof against cigarettes was nonetheless overwhelming.

In various ways it had been proven that smoking was harmful. Epidemiological studies had shown that smokers were more likely to develop lung cancer; animals grew tumours when their skins were painted with tar; pathologists had found harmful effects of smoking at a cellular level; and it had been demonstrated that cigarette smoke contained chemical substances that caused cancer. If that wasn't enough, all these studies were repeated, and each time they came to the same findings. A few years after it was published, Doll and Bradford Hill's study from 1952, for instance, had been rerun several times by researchers from Japan, the United States, Canada and France, and each time the outcome was the same: lung cancer patients tend to be smokers.[44]

At some point the evidence is so strong that even if one study gives the opposite result, the conclusion still stands. You see the same in research into climate change. One mild winter does not prove global warming is happening, but countless studies into coral reefs, glaciers, rises in CO_2, increasing temperatures do.[45] As was the case with smoking, these studies reach the same conclusion time and again. Researchers from different backgrounds, with different blind spots and interests, saw the same thing despite using different methods for measurement, data collection and analysis. If the evidence is that impressive, there is a case for 'scientific consensus'.

Such a consensus does not mean that 100 per cent of scientists side with a particular evaluation, nor that every study comes to the same conclusion. Science will never be able to give full certainty, because doubt is at its core. Knowledge has been growing for centuries because scientists have the courage to question the dogmas of their time. Nicolaus Copernicus had the courage to argue that the earth went around the sun, Albert Einstein dared to doubt Isaac Newton, and Archie Cochrane was cocksure enough to enter into battle with his fellow doctors.

But the tobacco industry used doubt – the core value of science – for its own benefit. Not to get closer to the truth, but to keep the public as far away from it as possible. It may have been scientists who lent it a hand, but it was also scientists who concluded at the end of the 1950s: we know enough. Cigarettes cause lung cancer.

The tobacco industry continued to deny the link between cigarettes and lung cancer for a long time. Until 1994, the bosses of the seven major cigarette brands claimed that they did not believe in the link. And as late as 1998, the then director of Philip Morris & Co. testified under oath: 'I do not believe that smoking cigarettes causes cancer.'

Yet inside the tobacco companies, things were quite different. As far back as 1953, nine months before the appearance of the mice study, Claude Teague – who worked at cigarette manufacturer R.J. Reynolds – had put together a survey of all the existing scientific studies into smoking.[46] His overview would ultimately serve as an exhibit in court cases against the tobacco industry, because it proved that manufacturers were already aware of harmful effects at an early stage. But Teague's report did not see the light of day until the 1990s, because – no surprise, really – it was never published.

How to lie with smoking statistics

To this day, the tobacco industry finances science. In 2017, news emerged that Philip Morris International funded the Foundation for a Smoke-Free World to a tune of 80 million dollars annually. The World Health Organisation reacted fiercely: this was a manifest conflict of interest.[47]

Outside the tobacco industry, doubt has become a powerful weapon against scientifically proven correlations. In their book *Merchants of Doubt*, Naomi Oreskes and Erik Conway reveal that the same tricks are used around climate denial.[48] And in the case of the international dairy industry, research was funded to cast doubt on the adverse effects of milk fat.[49]

It's a matter of time before new industries start to apply the same strategies to protect their interests. After Big Tobacco and Big Oil, maybe it will be the turn of Big Tech to keep investigations into the harmful effects of technology under wraps. Politicians can likewise be economical with the truth. High-ranking American officials casually dismiss claims about climate change under the banner of 'sound science'.[50]

Why did Huff and Fisher not know any better? Why did they continue to cast doubt on research about smoking and lung cancer? Perhaps Huff had been so used to cutting down research that he simply could not admit it when it proved to be sound. And perhaps statistician Fisher, a fervent pipe smoker, listened to his gut feeling when he criticised the tobacco research.

But there is a much more likely explanation. Fellow academic David Daube revealed how Fisher explained, not long before he died, why he had defended the tobacco industry: 'For the money.'[51] Huff, too, was paid by the tobacco industry. He had even been asked to write a book, which in the end never appeared. Its title? *How to Lie with Smoking Statistics*.[52]

CHAPTER 5

WE SHOULD NOT BE TOO FIXATED ON NUMBERS IN THE FUTURE

Meet sixty-five-year-old Jenipher.[1] For years, the Kenyan market trader earned her money selling food in the business district of Nairobi. Trade on her stall was brisk, but she barely had any money. She was not able to invest in her business, and if her health failed, she would be in immediate financial difficulty.

So what was the issue? For Jenipher, it was virtually impossible to borrow money. The sums she was able to raise through microfinancing were too small, the interest from loan sharks too high. And a normal bank would be unwilling to give her a loan, because she had no collateral. Furthermore, she was lacking something that was perfectly standard in other countries: a credit score.[2]

Credit scores have been a common phenomenon for decades in the Western World. In 1956, engineer Bill Fair and mathematician Earl Isaac set up their firm Fair, Isaac and Company (FICO) in the United States. FICO was based on a simple idea: with data at your fingertips you can better assess whether people will pay back their loans.

Until that point, the decision about whether you would be granted a loan was based on what people said about you, how you came across in the meeting and the banker's gut feeling about your trustworthiness. For many people this did not work out very well. In old American credit reports you can read how a particular liquor store is regarded as 'a low Negro shop' and that 'prudence in large transactions with all Jews should be used'.[3]

Fair and Isaac came up with a formula that looked at your finances, rather than your background. How much do you earn? Do you pay your bills on time? How much money have you already borrowed? On the basis of this data, they calculated a score that indicated the likelihood of you paying back your loan.

The FICO score proved to be a godsend for both sides: millions of people gained access to loans, and lenders earned more money, because the score was much better at predicting who would be defaulters than they were themselves. A formula, it transpired, led to better decisions than human judgement.

Credit scores are now used in many other countries. Nonetheless, millions of people do not have one yet. People such as Jenipher. However, in the last few years she has been able to obtain a credit score, as Shivani Siroya explained in a TED talk. Siroya is the CEO of Tala, a start-up that uses big data to grant loans. Jenipher may not have had a credit score, but she did have a mobile phone that tracked all kinds of data about her – her location, whom she texted, how long she was on the phone for, and so forth.

One day, Jenipher's son convinced her to install the Tala app. She applied for a loan and, in no time at all, was awarded one. Two years later her life has been transformed; she now runs three stalls and has plans for a restaurant. She can even apply now for a loan at a bank, as she has proven that she can manage her money.

One of the most dangerous ideas of our time

Jenipher's story is heart-warming. And even if it's a promotional story for Tala, it is telling about today's biggest-growing trend: the big data revolution. What makes data 'big'? Big data is often defined by the four Vs: Volume, Velocity, Variety and Veracity. In other words, vast amounts of different kinds of data moving swiftly and reliably.

The biggest difference between the current hunger for data and its equivalent from the time of Florence Nightingale – the 'first wave of big data' – is that these days we have the internet. The use of numbers still involves standardising, collecting and analysing, but because of the internet, it's happening on steroids. We *standardise* more than ever – from steps to clicks, from facial recognition to noise pollution.[4] We *collect* more than ever – per minute, Google performs more than 3.6 million searches, YouTube plays more than 4 million videos, and Instagram users post almost 50,000 photos on the platform.[5] And we *analyse* this mountain of data with increasingly clever methodology.

Along with the expansion of data, expectations about what we can do with it inflate. Tala, the company that gave Jenipher a loan, wants to use big data to reach the people who currently do not have access to credit. The American counselling service Crisis Text Line analyses the data of text messages to identify people vulnerable to suicide.[6] And the organisation Rainforest Connection collects data with old smartphones to combat illegal logging and animal poaching.

Expectations are sky-high. Policy-makers, company executives and public intellectuals argue that we can solve the climate crisis with big data,[7] transform health care[8] and eradicate hunger.[9]

We might even save democracy with big data. Elections are pointless if a large number of people do not vote, university administrator Louise Fresco argued in 2016 in an opinion piece in the Dutch newspaper *NRC*. 'What if we replace democratic elections with a system of artificial intelligence?'[10] Clever calculation systems could render elections superfluous, because our preferences are already stored in our data – where we travel, whom we talk to, what we read.

All that information about our behaviour – enhanced, if necessary, by additional surveys – could be used to distil what we really consider important and thus our political preferences.

Fresco's thought experiment may seem completely outlandish, but make no mistake: big data algorithms are already becoming ever more powerful. Insurers use them to calculate what your premium should be,[11] tax authorities to work out whether you will commit fraud,[12] and American judges to assess whether a prisoner should be released early.[13] More and more, our fate is in the hands of big data. The assumption that we can be oblivious and let numbers make decisions about our life is dangerous. Behind this notion lurks a serious misunderstanding: namely that the data always corresponds with the truth. That, with big data, the problems we have seen in the previous chapters cease to exist.

It's high time we looked at big data a bit more closely, through the lens of the previous chapters. How are we standardising, collecting and analysing in the twenty-first century? And why can we not leave important decisions to numbers and calculation methods without giving it any further thought?

What we talk about when we talk about algorithms

Let's begin with a look under the bonnet. How is data now being used? Just as in the past, averages and graphics were invented to understand the mountains of information that existed at the time, clever people today come up with methods to tame trillions of bytes of data. These techniques – algorithms – decide which search results you get on Google, which posts you see on Facebook, who pops up on your dating app and who receives a loan from companies such as Tala. (The word algorithm is derived from the ninth-century Persian mathematician Muhammed ibn Musa al-Khwarizmi, who wrote a book on algebra.)[14]

In fact, an algorithm is nothing more than a number of steps you take to reach a particular goal. On a computer screen it looks very dry and dull: line by line, a software developer instructs in computer language which steps have to be taken under which circumstances. Such a line can be an 'if–then command', for instance: '*If* someone has paid back her loan, *then* her credit score goes up by ten points.'

How does an algorithm work? In her book *Weapons of Math Destruction*, the American mathematician and author Cathy O'Neil explains it using a practical example: cooking for her family.[15] She is happy when her family (a) eats enough, (b) enjoys the food and (c) gets sufficient nutrients. By evaluating these three factors every night, she knows how dinner time went and how it can be improved. The observation that her children do not touch spinach but devour broccoli helps her to make them eat a healthier diet. A few constraints have to be taken into account for her to achieve her objectives, however. Her husband is not allowed to have salt and one of her sons does not like hamburgers (but loves chicken). Nor does she have unlimited budget, time or appetite for cooking.

Following years of practice, O'Neil has become very adept at the process. She has, in part subconsciously, developed an ever-tighter step-by-step plan to serve up the best meal for her family. Now, suppose a computer took over that task. How would she be able to transmit her menu decisions to this machine? She could begin with thinking up a way to *standardise* her objectives. In order to find out whether her family eats enough tasty and healthy food, she could, for example, look at (a) number of calories, (b) satisfaction scores and (c) percentage of the daily recommended amounts of each nutrient. She should also work out how to quantify the constraints, for example by setting an upper limit for her budget.

Once she has worked out what and how she will standardise, she can start to *collect* data. She could first draw up a list of possible recipes, including preparation time, price and nutritional value. Per meal, she can note down how this food scores in terms of quantity and healthiness, and she can ask the members of her family to award each meal a score between one and ten.

With this data, O'Neil could write a programme that spells out exactly what her family should eat each day. But she can also set up a programme that is self-learning. As long as everything is cast in numbers, the computer itself can *analyse* the correlation between the meals and the objectives. And maybe the algorithm will pick up on patterns that she herself had never noticed before; for instance, that her children can stomach more sprouts if they eat pancakes the previous day. The computer thus uses machine learning, a type of artificial intelligence, to learn a task that has not been pre-programmed step by step.[16] The eerie thing about this is that, because of the programme's self-learning capabilities, algorithms can become so complex that no one, not even the programmers, understands which steps the software is taking.

In short, O'Neil would have standardised her cooking task, collected data and made the software analyse it. Where have we seen these steps before? They are exactly the same steps we saw Florence Nightingale, Archie Cochrane and others take. And with algorithms, as in the previous chapters, a great deal can go wrong during each of these three phases.

1. The trouble with measuring abstract concepts

In the financial sector, there are companies like Tala which use big data to assess someone's creditworthiness. Take ZestFinance, which has been giving more than 300 million individuals a credit score since 2009. The company, set up by former Google CIO Douglas

Merrill, argues that the traditional credit score system is hampered by 'little data'.[17] Conventional credit scores, as devised by Fair and Isaac way back in the past, use 'less than fifty data points', which is 'only a fraction of the public data available on any given person'. Zest, conversely, uses more than three thousand variables in order to assess someone.[18]

In the Netherlands, too, countless companies are using big data in order to measure clients' attitudes towards payment. Dutch data trader Focum gives a score of between one and eleven.[19] Haven't paid your bill yet? You lose ten points, whether the sum involved amounted to 20 or 20,000 euros. Such credit raters sell the scores to whomsoever is willing to buy them, from insurers to housing corporations, from Vattenfall to Vodafone. A bad credit score can mean that you'll be refused a phone contract or might have to pay a large deposit when you sign up with a new energy supplier. The company claims to possess data for 10.5 million Dutch people. That's a lot for a country of just 17 million.

You might wonder, what's wrong with all of this? Credit scores also offer opportunities after all, as the story of Jenipher from Kenya demonstrated. Yet credit scores can have a greater impact on your life than you might think, and not necessarily always a positive one.

We saw earlier that an IQ score is an *approximation* of something as intangible as intelligence. The same goes for credit scores. These scores try to convey how likely it is that you will pay back your loan in the future. A credit score, in other words, is a prediction.

Many big data models try to predict the future. In the American criminal justice system, a calculation is made of an offender's likelihood to re-offend. These calculations have serious consequences; they play a part in the decision as to whether someone will be granted an early release or not.[20] But if there is one thing that is abstract and difficult to predict, it's what will happen in the future. Statistical models behind these kinds of predictions are never watertight; they always involve a considerable degree of uncertainty. When we forget that such predictions are only an *approximation* of someone's behaviour, we judge people on the basis of inadequate data.

There's another issue with credit scores. They tend to be used to express something other than future behaviour, something that is as least as abstract: trustworthiness. These scores are not just being used for granting loans. On American dating site CreditScoreDating.com – 'Where Good Credit is Sexy' – you can look for someone who is a credit-score match for you.

But the use of credit information goes even further. An American study among HR professionals from 2012 showed that some 47 per cent of employers check the credit history of job applicants.[21] Another study among American households with credit card debt noted that one in seven respondents with a bad credit history were informed it was the reason they were turned down for a job.[22]

These findings apply to specific samples and are therefore not representative for the entire American population. But it's an undeniable fact that employers check the background of their job applicants. One glance at American online vacancies reveals that employers demand credit checks for jobs as diverse as selling fireworks and assessing insurance claims.[23]

Employers do not get to see a credit score, but receive a credit *report*, an overview of someone's borrowing behaviour. With this data, employers hope to be able to assess the character of a potential employee, and whether they will commit fraud in the future.[24]

Yet there is no evidence whatsoever for a link between your borrowing behaviour and your performance on the shop floor. The few studies that exist do not show a correlation. Researcher Jeremy Berneth and colleagues compared individual FICO scores with personality tests.²⁵ People with higher credit scores performed better in conscientiousness tests, but were less agreeable than those with lower scores. For other characteristics there was no difference.

More importantly, there was no relationship between credit scores and fraudulent practice. In sum, it is wrong to use someone's credit history as an approximation for trustworthiness at work. For good reason, it's now unlawful for an employer to ask for someone's credit history in eleven American States.²⁶

But even if your credit data is used exclusively for granting a loan, we should still be on our guard. Because in the collection of data, big or otherwise, many things can go awry.

2. Big data can have murky origins

Big data can help solve fundamental problems in data collection. As the name implies, sample size is no longer an issue. Almost everyone is on the internet. Besides, various appliances and devices – thermostats, cars, Fitbits – track what we do. And cities such as Dubai, Moscow and New York call themselves smart cities because they gather all kinds of data about their citizens with new technology, from WIFI-trackers in lamp posts to sensors in fibre-optic cables.

Since we now use more technological gadgets in our day-to-day lives, there's no longer much need to conduct personal interviews, as sexologist Alfred Kinsey did in his study. Now you can observe directly what people are up to. As data researcher Seth Stephens-Davidowitz puts it: 'Google is a digital truth serum.'²⁷

Married women, for example, ask Google eight times more often whether their husband is homosexual than whether he is an alcoholic; in India 'my husband wants' is followed most frequently by 'me to breastfeed him'; and even though men from conservative states such as Mississippi report less often in surveys that they are homosexual, there are relatively just as many searches for gay porn there as in progressive states such as New York.[28] Alfred Kinsey would have had a field day with this data.

The companies behind credit scores know that, in this information age, personal data is there for the taking. They no longer need to request it via official routes, but can comb the internet for information about you. As CEO Douglas Merrill of ZestFinance puts it: 'All data is credit data.'[29] Sometimes the data they collect is public, such as registration information at the Chamber of Commerce, but at other times you – often without your knowing it – have given permission for your data to be shared.

Not infrequently, data emerges from more obscure corners. In October 2017, the Dutch *Groene Amsterdammer* weekly and the Investico platform published a thorough investigation by journalists Karlijn Kuijpers, Thomas Muntz and Tim Staal into data traders in the Netherlands.[30] They discovered that some bureaus had received data directly from debt collection agencies. The financial history of these people ended up in a database, unbeknownst to them, and they continued to be blacklisted as a result – long after they had paid off their debts. This practice is illegal, by the way, because you should be informed if your data is shared with others.

It's often impossible to find out whether the data used is correct, because it's unclear which data has been used in the first place. The three Investico journalists found that a housing corporation from the Dutch town of Wageningen can refuse people access to social housing if their credit score is too low, but that the organisation 'does not need to know how the data firm arrives at these scores'. To put it to the test, the journalists asked ten people to request their own data from three data bureaus. The results were paltry; they received close to nothing. But when the journalists pretended to be a commercial client and bought the data about the same people, they suddenly received extensive data reports.

It's beyond dispute that there are frequent errors in data. The American Federal Trade Commission noted in 2012 that, in its sample, a staggering quarter of all people found errors in their credit reports from one of the three big bureaus.[31] For one in twenty, the divergence was so great that these people may have wrongfully had to pay a higher interest rate for loans.

Such mistakes crop up in other databases as well. Between 2009 and 2010 there appeared to be 17,000 pregnant men in the UK. That's right, pregnant *men*. The code that registered their medical treatment had been mixed up with that of an obstetrics procedure.[32] Data errors occur everywhere: wrong addresses in the municipal personal records database, incorrect income figures lodged with the tax authorities and the employee insurance agency, wrongful registration as a criminal in a police database. Not a good idea, therefore, to put blind trust in numbers.

Occasionally mistakes come about as a result not of bungling, but of mal-intent. Equifax, one of America's biggest credit bureaus, announced in 2017 that it had been hacked. The data of almost 150 million consumers – almost half of the American population – had been stolen.[33] Names, dates of birth, addresses and Social Security Numbers could now be sold on the black market. And these details were valuable, because with them you can carry out practically every important transaction in America. You can apply for a credit card, file your tax return and even buy a house on behalf of someone else. Needless to say, the resulting data doesn't say much about the people whose information was stolen.

As an old statistics adage goes: 'Rubbish in, rubbish out'. You can build the smoothest machine-learning algorithm, but it will be of no use to you if the data is flawed. Suppose that, in the future, data fraud has been eliminated and we have flawless data at our disposal: will we then be able to leave fate to algorithms?

3. Correlation is still not the same as causality

A traditional credit score, like the FICO score, is based solely on data about you. Whether you have borrowed money, how much you have borrowed and whether you paid it back in time. The idea is that these factors can predict whether you will pay off your loan in the future.

There is a good case to consider this reasoning inequitable. Debts are frequently caused by high medical costs or job losses. Some people are able to absorb these setbacks with their savings, but not everyone has enough capital to do so. A credit score is thus a measure not only of trustworthiness, but also of sheer luck.[34]

The calculation of big data credit scores goes a step further. Back to Jenipher and her food stall. How did Tala decide that she was allowed to get a loan? Jenipher had to give the company access to her phone via an app, which holds a goldmine of data waiting to be analysed. Her location history revealed that she was often on the move, but with a regular pattern. She was either at home or at her stall. Her telephone data showed that she was frequently on the phone to family in Uganda. Aside from that, she communicated with no fewer than eighty-nine different people.

Each of these are factors that according to Tala's algorithm increase Jenipher's chances of her paying back the loan. The fact that she has regular contact with loved ones increases this chance by 4 per cent, for instance. A fixed daily pattern and having more than fifty-eight contacts likewise appear to be positive signs.

Jenipher's example shows how big data credit scores work differently from traditional scores. The algorithms look not only at what *you* have done, but at what *people like you* have done. They look for links – correlations – in the data and predict what you will do. All numbers are welcome here, as long as they predict correctly.

Even the words on someone's application form can be telling. ZestFinance's Douglas Merrill suggested in 2013 that an application completed in capital letters – or only in lower case – could be an indication of poor payment behaviour.[35]

Shopping habits, too, can indicate whether someone will pay back his or her loan. In 2008, American Express decided to close down the credit cards of some of its American clients.[36] 'Other customers who have used their credit card at establishments where you recently shopped have a poor repayment history with American Express', the company wrote. American Express later denied that it had blacklisted particular shops, but it did admit using 'hundreds of data points' in order to monitor creditworthiness.

Another data goldmine: social media. In 2015 Facebook secured a patent to use your social network for calculating credit scores.[37] The rationale behind this? If your friends have a bad credit history, you too can probably not be trusted with a loan. NEO Finance already uses LinkedIn data to assess someone's 'character and capacity' by checking whether his or her CV conveys the truth.[38]

There was a time when bankers allowed their decisions to be influenced by prejudices about race, gender and class. The FICO scores were supposed to end this. But with big data credit scores we seem to be doing exactly the same as the bankers of yore: judging someone on the basis of the group he or she belongs to.

It's just that these groups are now defined as Capital Letter Writers, Bargain Hunters, the Friendless. Look below the surface and you will see little that's new. Writing in upper case is probably related to your level of education. Having LinkedIn contacts to having a job. And where you shop says a great deal about your income. Algorithms are thus making exactly the same distinctions as that old-fashioned banker; poor or rich, employed or unemployed, highly educated or poorly qualified.

Statisticians call them correlations, other people prejudices.

*

So where are we with correlation and causality now that we have big data? According to Chris Anderson, former editor-in-chief of technology magazine *Wired*, we do not need to worry about that. The explanation for particular relationships is unimportant, he wrote in 2008 in his influential article 'The End of Theory'.[39] 'Google's founding philosophy is that we don't know why this page is better than that one: If the statistics [...] say it is, that's good enough.' That correlation is not in fact the same as causality, as we saw with storks and babies, no longer matters according to Anderson. 'Petabytes allow us to say: "Correlation is enough."'

It's an extremely naïve statement. In the era of big data, correlation is still not enough. Take Google Flu Trends, the algorithm that was launched with much fanfare in 2008.[40] Google promised that, using searches, it would be able to predict when, where and how many cases of flu there would be. The idea was that if people were ill, they would google the symptoms.

The promise was great. Then Google CEO Eric Schmidt argued that tens of thousands of human lives could be saved each year.[41] And at first he appeared to be proven right. For two or three years the model predicted fairly accurately when and where flu would strike. But in the years that followed, the algorithm got it wrong each time, with an absolute low point in 2013 when it predicted double the actual incidences of flu.[42]

Where did things go wrong? The creators of the algorithm had selected forty-five search terms – out of fifty million – which correlated most strongly with the progress of the flu outbreak. They then tracked the searches for these terms. This sounds logical but, as is the case with smaller data sets, the jelly bean problem lurks; if you look long enough, you will always find a correlation.

Worse still, big data makes this problem even more of an issue, because the more variables you have, the more correlations you will find that are significant. Simply by chance. The researchers, for example, found a strong correlation between the search term 'high school basketball' and the spread of flu.[43] They manually removed these kinds of spurious correlations from the model. But this is not always an easy decision, because how do you determine whether something is coincidence? Is the search term 'handkerchiefs' a coincidence because it's winter, or does it signify a flu outbreak?

Another problem with the algorithm was that the designers ignored important developments, such as changes in the design of Google's own search engine. From 2012, the website showed possible diagnoses if someone looked up 'coughing' or 'fever', for instance. One of the diagnoses? Flu. This led to people starting to search for information about the illness which meant that the Google Flu algorithm overestimated the flu outbreak.

We saw earlier that credit bureaus also make predictions, just like Google Flu Trends. Spurious correlations also lurk in these predictions, and important developments can likewise throw a spanner in the works. For instance, once it becomes common knowledge that you have to use particular words in an application form, people can game the system, rendering the correlations meaningless.

But suppose that, in the future, we no longer need to worry about these two pitfalls. That we find ways to recognise spurious correlations and monitor changes in real time. This will still leave a problem that cannot be solved, because the way that we use scores affects what the scores look like.

The numbers that should have captured reality have replaced it

> 'I'm not investing because you won't hire me.'
> 'I didn't hire you because you weren't investing.'

In 2003, this exchange took place in the American state of Virginia.[44] It might have been a fiery exchange between an employer and a job-seeker. Perhaps the job-seeker was turned down on the basis of his skin colour. Or perhaps the employer had cast one glance at his CV and decided: not well-educated enough.

But the applicant was not black, he was purple. And the two were not a real job-seeker and employer, but students. They were taking part in an experiment conducted by Harvard professor Roland Fryer and colleagues. Their study would show how quickly an equal world can be derailed when you focus exclusively on numbers.

In the experiment students were randomly assigned a role as 'employer', 'green job-seeker' or 'purple job-seeker'. During each round, job-seekers had to choose whether they would invest in their own education or not.

On the one hand, there was something to be said against an investment of this kind; the students were paid a fee for taking part and 'education' would cost them money. On the other hand, their chances of achieving a higher score in the 'test' (which consisted of a kind of weighted dice that would roll more often in their favour if they had invested in education) went up, increasing their chances of earning extra cash. Employers were keenest on applicants with a good score, because an educated employee brought in more money. But because an employer would only ever see the test score, they were never 100 per cent certain whether the applicant had actually been educated. The experiment is very similar to reality; an employer never knows for sure whether an applicant is suitable, but they can make an assessment based on imperfect criteria such as exam marks.

In the first round of the experiment, purple applicants invested a little less money in education. This had nothing to do with their purple identity, because the colour had been assigned randomly. During the following round, employers were able to see the statistics.

They thought they'd be better off without the purple workers. When the purples for their part saw that their green colleagues were hired more often, they decided to invest less because their investment did not seem to increase their chances of being offered a job.

The curious thing was that everyone behaved in a rational manner. Judging by the numbers, this seemed the best strategy. But within twenty rounds a vicious circle emerged, which resulted in an extremely unequal world. 'I was amazed. The kids were really angry,' Fryer told Tim Harford, who wrote about the experiment in his book *The Logic of Life*. 'The initial asymmetries came about because of chance, but people would hang on to them and wouldn't let go.'

Obviously, the world is a great deal more complex than in this fascinating experiment. But it does illustrate a powerful message: numbers are both the cause and effect of what the world looks like. They may seem to be passive registrations of reality, but nothing is further from the truth: they shape reality. And the more numbers rule our world, as is now happening with big data, the more they will change it.

Take 'predictive policing', algorithms that are used by the police to find out who might be a criminal. American data shows a clear correlation between poor black young men and criminality. On the basis of these algorithms a police force would want to focus on neighbourhoods and individuals that meet this description. The result? Racial profiling, with the consequence that many innocent people are also getting arrested. And when you arrest particular people more frequently, they will automatically end up in the statistics more frequently as well. You will overlook the rich white criminals, because they fall outside your operational remit. So it's no surprise then that in the subsequent statistics you will see – perhaps even more strongly – a link between skin colour and criminality.

You will run the same risks with credit scores; people with

particular characteristics find it more difficult to get loans than others, landing these people in poverty more quickly, making it even harder for them to get a loan, which accelerates their poverty, et cetera, et cetera. Algorithms like this become self-fulfilling prophecies.

The numbers that should have captured reality have *replaced* it.

What do you want to achieve with numbers?

In 2014, the Chinese government announced that, from 2020, there would be a country-wide introduction of a 'social credit system'. According to China's leaders, this is essential for 'building a harmonious socialist society'.[45] The score system will 'allow the trustworthy to roam everywhere under heaven while making it hard for the discredited to take a single step'. During the past few years we have been able to catch a glimpse of the system, because in 2015 the Central Bank of China selected eight companies to experiment with trials.[46]

One of those companies is Ant Financial, the Chinese company behind Alipay, the payment app for the omnipotent web shop Alibaba. The app has more than half a billion Chinese users[47] and offers pretty much every service: payment in shops, buying train tickets, ordering food, calling a taxi, borrowing money, settling bills, paying fines and making friends. It's as if your bank app has merged with Amazon, Facebook, Uber and your Oyster card. And since the Central Bank's decree, a new service has been added: Sesame Credit, a points system that gives you all kinds of benefits.

On Sesame Credit, participants are given a score between 350 and 950 points.[48] If your score is above 600, you are allowed around 600 euros credit for the Alibaba online shop. If you have more than 650 you do not need to pay a deposit when you hire a car. And a rating of 700 points or higher makes it easier to apply for a visa. A higher score is also good for your reputation; you can use it to brag on social media and it will give you a prominent spot on dating sites. Sesame Credit, as the name implies, opens doors.

How can you accrue points? You have to pay your bills on time, not miss out on a month's rent, and pay off your loans. If you have filled in your personal details – your address, job, qualifications – you get a higher score. And what about the purchases you make via the app? Ordering too many games is bad for your score, Ant Financial's Technology Director explained in an interview with *Wired*, but buying nappies will generate more points. This statement was later disowned by the company, but it does make you think. There's no end to the score system's possibilities if you realise which data can be collected through the Alipay app.

On top of this, Sesame Credit uses data from other sources. If you have cheated on a test, woe betide you. Sesame Credit's General Director stated in 2015 that she would like to have a list of students who had cheated during their national entrance exam, in order to punish them for their 'dishonest behaviour'. And the company has used a government blacklist, containing millions of people who had not paid their court fines, in order to downgrade the scores of defaulters.

Big data is intimidating. The scale is unprecedented and the algorithms are sometimes so complex that even the developers cannot make head or tail of them. But in the end, big data is about the same question as small data: what do you want to achieve with the numbers? China may be unequivocally clear about the objective of the social credit system – 'building a harmonious socialist society' – but we must be aware that every single algorithm is laced with moral choices.

Each algorithm tries to optimise something. YouTube, for instance, wants you to carry on watching for a long as possible, because that brings in money via advertisements.⁴⁹ Whether a clip is truthful is of less importance. Guillaume Chaslot, former Google engineer and founder of the AlgoTransparency website, began to delve into YouTube's algorithms. He discovered that the platform recommended videos describing the earth as flat or revealing that Michelle Obama is a man. 'On YouTube, fiction is outperforming reality,' Chaslot told the *Guardian*.

The police likewise, when they use a predictive policing algorithm, try to optimise something, namely our security. But this objective is at odds with another: justice. Is it justifiable that innocent people are arrested? It depends on what outcome you want to achieve.

The same goes for credit scores. Earlier in this chapter we saw the Federal Trade Commission's conclusion that one in twenty credit reports contains serious errors. The Consumer Data Industry Association (CDIA), the professional association of credit bureaus among others, celebrated this as a positive message. After all, 95 per cent of consumers were not affected by mistakes.⁵⁰

But is 5 per cent a lot or a little? It depends on what you intend to do with the scores. Money lenders tend to be commercial parties. Their objective is generating profit. Viewed through their lens, 95 per cent is perfectly respectable. Whether it's fair is less important to them. The borrower is not the client, but the product.

We have to remain vigilant. The idea of introducing a social credit system may seem a ruthless instrument from an autocratic regime, but in the UK and other countries we are also extensively scored. In the words of technology journalists Maurits Martijn and Dimitri Tokmetzis: we live in a 'scoreboard society'.⁵¹

A credit assessor tries to calculate whether we can handle money, an insurer whether we will stay healthy, the tax authorities whether we will commit fraud and the police whether we will break the law. Each time, these calculations have consequences for our daily life: you are refused a loan, you receive a collection letter, you are arrested, you have to pay a higher premium. And often it's the people already occupying a vulnerable position in society who are the ones that are hurt the most.

Big data can make the world better. Just look at Jenipher in Nairobi, who was able to improve her life thanks to a loan. But these same algorithms that can help people like Jenipher have the ability to maintain centuries-old inequalities as well as create new ones.

So it's not the algorithm itself that is 'good' or 'bad', but the way we use it. That's why it's vital to join in the discussion about the question of what purpose algorithms serve. Is our aim to find the truth or generate profit? To prioritise security or freedom? Justice or efficiency? These are moral dilemmas, not statistical ones.

Algorithms will never be objective, however reliable the data may be and however advanced artificial intelligence becomes. When we forget this concern, we leave the moral decisions to people who happen to have a talent for computers. And, while they are programming, they will decide what is good and what is bad.

CHAPTER 6

OUR PSYCHOLOGY DECIDES THE VALUE OF NUMBERS

'One glass of alcohol is one too many.' I saw this headline flash past me on the website of the Dutch national broadcaster NOS in April 2018.[1] Drinking more than one glass of alcohol a day increases your chances of dying early, the report stated.[2]

The article referred to a study published in the renowned publication *The Lancet*, for which eighty-three studies referring to a total of some 600,000 study subjects had been aggregated.[3] Impressive, I thought, but correlation is not the same as causality.

The same thing was spotted by Vinay Prasad. Prasad, an American doctor and researcher, who knows everything there is to know about evidence-based medicine, had delved into *The Lancet* study and tweeted gruffly: 'A team of scientists prove [sic] the human thirst for bullshit science and medicine news is unquenchable.'[4]

He then elucidated his statement in a thread of more than thirty tweets. He mentioned publication bias that we've seen in earlier chapters. He also argued that, in this study, alcohol use had only been monitored for a short time And although a high mortality risk had been found among beer drinkers, in wine drinkers it turned out to be minimal. It wasn't so much the alcohol, Prasad suggested, but the lower income of beer drinkers that was unhealthy.

I came to the conclusion that there was nothing wrong with a tipple or two.

Why do things keep going wrong?

When I was writing my first articles as the numeracy editor of online journalistic platform the *Correspondent*, I thought I knew the solution to the dogged issue of number misuse: more knowledge. According to the Organisation for Economic Cooperation and Development (OECD), one in four adults in developed countries perform at or below the lowest level of numeracy – they find it difficult to interpret statistics and diagrams.[5] Mathematics anxiety is such a serious phenomenon, the OECD concluded in 2012, that it occurs in around 30 per cent of fifteen-year-olds.[6]

If only news consumers could understand how numbers work, I thought, everyone would automatically spot their misuse. So I began to write about bad polls, about margins of error, about correlation and causality. And, each time, I tried to explain how to recognise these kinds of errors, to prevent future misunderstandings.

More knowledge as the ultimate solution; it seems so logical. When climate scientists publish temperature graphs, when journalists fact-check political statements, when politicians trot out economic figures in a debate – each time they try to combat mistakes with ever more information.

But the longer I wrote about number misuse, the more I began to doubt whether knowledge was the only solution. I stood in a long line of writers who have wanted to raise awareness about this topic, but little seemed to have changed. Darrell Huff had already outlined the principal pitfalls with numbers more than sixty years ago, in *How to Lie with Statistics*. The book had been a bestseller, but the same mistakes are still being made today. The discussion about IQ and skin colour crops up in every new generation, unrepresentative polls continue to get far too much attention, and health news that confuses correlation with causality flashes by almost daily.

It's often easy to recognise these errors by asking a few questions. How was the data standardised? How have the figures been collected? Is there a causal relationship? I have discussed these questions extensively in the previous chapters and list them once more at the end of this book.

Yet erroneous conclusions about numbers keep slipping past the scientists, journalists, politicians and newspaper readers. And past me. I wished the ground would open up and swallow me when, after a lecture, I saw that 50 per cent of those attending had not rated my performance as good. But I forgot to take into account that only two people had taken part in the survey.[7] And I was outraged when I read about a study that alleged that female programmers were undervalued by their colleagues. Later it turned out that the media had misinterpreted the study; programmers were not nearly as sexist as the reporting had suggested.[8]

Over and over again, I fell for the same mistakes I discussed at length in my articles. It was only when I started working on this book that I really began to understand why this was happening. When it comes to numbers, the issue is not only errors in reasoning, as I had thought, but also gut feelings. In numerous instances in this book, researchers were influenced by their – conscious or unconscious – biases and convictions.

And we number consumers are equally prone to this.

An interpretation that isn't good, but feels good

For years, Yale professor Dan Kahan has been investigating how culture, values and beliefs affect your thinking. In one of his experiments he and his colleagues presented participants with a table showing the results from a fictitious skin cream trial.[9] In one group, the figures showed an increase in skin rashes; in the other, they had decreased. Does the cream help with the rash, Kahan asked, or does it make it worse?

To find the answer, participants had to make a tricky calculation with the figures shown in the tables. The people who had scored better in an earlier maths test tended to come up with the right answer. Until this point the experiment confirmed what you'd expect: that if you have a better understanding of numbers, you will get closer to the truth.

But there were two further groups of test subjects. They were given the same tables of figures, but this time representing a controversial topic in American politics and media: gun control. It featured a fictional experiment with stricter legislation. The question this time was: does crime go up or down as a result of the new measure?

The answers were as different as day and night from the answers given by the participants in the skin cream 'experiment'. Those who were good at maths performed worse than before. The figures were exactly the same as those presented for the skin cream experiment, but now the participants gave the wrong answers.

The explanation for Kahan's results? Ideology.[10] Irrespective of the actual figures, Democrats who identified as liberal, normally in favour of gun control, tended to find that stricter laws brought crime down. For the conservative Republican participants, the reverse was the case. They found that stricter gun control legislation did not work.

These answers are no longer to do with the truth, Kahan argued. They are about protecting your identity or belonging to your 'tribe'. And the people who were good at maths, Kahan also found, were all the better at this. Often completely subconsciously, by the way. It was their psyche that played tricks on them.

Time and again, Kahan saw this result in his experiments; when people know more facts or have more skills, they have more to choose from while deluding themselves.[11] Our brain works like a lawyer; it will find arguments to defend our convictions, whatever the cost.

This can even mean that you believe one thing at one point and something else later on. There are American conservative farmers who deny the existence of climate change, for instance, but who take all kinds of measures to protect their business against the effects of a changing climate.[12] This seems irrational, but it isn't, Kahan explains. Much can be at stake if you alter your convictions. The farmer who suddenly believes in climate change is given the cold shoulder by his family, in church, at the baseball club. He puts a great deal on the line but gets nothing in return. It isn't as if he's going to change the climate on his own. The truth will have to wait.

Everyone is susceptible to these kinds of psychological pressures, including Kahan himself. In an interview with journalist Ezra Klein in 2014 he mentioned that he always assumes he will make the same mistakes as the ones he observes in his research.[13] And he, also, protects his identity with 'facts'. In short, a good interpretation of the figures is not just about what we know, but also about our psyche. So how can you be mindful of your own biases when you come across numbers? Here are three tips.

1. What do you feel?

There are plenty of issues in which the psychological processes from Kahan's study do not play a part. Most people will have a neutral reaction to numbers concerning something like skin cream. But it's the numbers about which you *do* feel something that are susceptible to bias. Racism, sex, addictive substances – the chapters in this book deal with such controversial issues for good reason. They are issues that are closely related to your identity and 'tribe'.

Should you just eliminate these feelings? That would be impossible; they are there, whether you like it or not. And this is a good thing. Without fear we would blindly walk into dangerous situations. Without anger we would not stand up against injustice. And without joy, life would be soulless. Feelings are part of us.

So when you see a number, take a step back and ask yourself: what do I feel? When I saw the alcohol study mentioned above, for example, I became irritated. Especially when I later read the headline 'An extra glass of alcohol can shorten your life by 30 minutes'.[14] This was simply total nonsense. My irritation was a feeling that matched my professional 'tribe' – number sceptics – but also my personal one. When I meet friends, we drink a few glasses of wine or beer. That's what we do. Should I stop doing this? I'd rather not. I felt pleased when I read the tweets from the renowned Vinay Prasad. Relieved; I could carry on drinking.

But I had overlooked an important factor. When I realised that I felt particularly upbeat at the conclusion there was nothing wrong with drinking, I had another look at Prasad's tweets. And I saw that nowhere had he said that drinking was *not* harmful, just that this study was flawed.

As in Kahan's study, I had immediately chosen an interpretation that fitted my 'tribe'. An interpretation that was not necessarily the right one, but that *felt* right. I was good at this kind of thing because, as a result of my work, I knew every argument against this type of study. My brain, too, had worked like a lawyer.

2. Go another click!
At the beginning of 2017 Dan Kahan and his colleagues published a new study.[15] He had asked around five thousand people questions to measure their 'science curiosity' for a project about science documentaries.[16] How often did the respondents read books about science?

Which topics interested them? Did they prefer to read articles about science or about sport?

He also asked a few questions about the respondents' political persuasion and their ideas about climate change. 'How much risk do you believe global warming poses to human health, safety, or prosperity?' was one of them. In the same way that Kahan had used a maths test in his earlier experiments, he was now measuring 'science intelligence' – a skill that was supposed to help with interpreting information about climate change.

Kahan again saw what he had found in earlier research: liberal Democrats saw more risk than conservative Republicans. And the more 'intelligent' the respondents, the bigger the difference between the two groups.

But what if he did not categorise according to intelligence but instead according to curiosity? These two were not the same, he saw in his data. Someone could be very curious about science, but not necessarily be any good at it – and vice versa. When he looked at the correlation between curiosity and the perceived climate change risk, he saw an interesting outcome: the Democrats and Republicans still had different opinions, but the more curious the subjects were, the greater they perceived the risk of the earth warming up. Irrespective of their political convictions.

Why did curiosity play a part here? In a follow-up experiment Kahan presented respondents with two articles about climate change; one that confirmed the concerns about it, another that was sceptical. The headline of one of the articles had been worded in such a way as to appear surprising: 'Scientists Report Surprising Evidence: Arctic Ice Melting Even Faster Than Expected'. The other article appeared to be reporting nothing that was new: 'Scientists Find Still More Evidence that Global Warming Slowed in the Last Decade'. Which article do you want to read, he asked? And this is where he discovered the power of curiosity. Curious types did not choose the article with the headline that accorded with their convictions, but the challenging one. For these respondents, curiosity was a stronger force than ideology.

This experiment is educational. If you encounter a number, don't stop and just accept it, but go and explore. Search – on- or offline – for people who look at the number from a different angle. Don't only read articles that confirm what you already think, but look for information that may make you feel uncomfortable, angry or desperate. As writer Tim Harford puts it: 'Go another click.'[17]

I put this to the test and began to search for more information about the impact of alcohol on our health. Some googling soon led me to all kinds of studies that suggested a causal link between alcohol and cancer risk. An experiment with a baboon that developed liver disease as a result of alcohol consumption,[18] for instance, and a meta-study that showed a linear correlation between breast cancer risk and alcohol consumption.[19]

What became clear to me is that experts have long agreed that drinking has adverse effects. Since 2015, the Dutch Health Council has recommended drinking no more than one glass of alcohol a day, for a reason.[20]

3. Accept uncertainty

Kahan's research into curiosity is still in its early stages. His experiments have to be repeated, and even if these replications show the same results, his conclusions may yet be invalidated by new studies.

Many of the figures you see in the newspapers are no different. They come from thorough, peer-reviewed research, but they are premature because still more research needs to be conducted. Should you ignore such inconclusive figures? No, like Kahan's studies, they help us to understand the world a little better. But do take them with a pinch of salt. And bear in mind that, in a few years' time, people may reach different conclusions.

Research into alcohol is much more advanced than Kahan's research into curiosity. When you start to investigate and google 'meta-study' (a study into studies), you soon see that many alcohol studies come to the same conclusion. The causal link between breast cancer and alcohol consumption has now been proven. Alcohol researchers came to the same conclusion as scientists who had examined the stacks of studies about the impacts of smoking: we know enough. But even if the research into alcohol is never definitive, that is the nature of science. There are studies that suggest that moderate alcohol intake even combats some diseases. Moreover, you cannot always disentangle correlation and causality in alcohol studies; research on animals is not the same as research on people, after all; and it's unclear how much alcohol you can drink before it becomes bad for you.

As it turns out, uncertainty is something we do not handle psychologically very well either. There is a reason why people with firm convictions dominate talk shows, political debates and newspaper columns. *I'm sure about this*, each and every one of them projects, *let me tell you how the world works.*

But people who are certain, by definition lack curiosity. If you hang on to your convictions at all cost, you are never receptive to new information. If we want to use numbers well – and information in general – then we have to embrace this uncertainty. I pointed this out earlier: numbers are a window onto reality, but the view they offer is no more focused than that seen through frosted glass. At best, all they show is the general outline.

Do not let yourself be paralysed though. At some point you will have to make a choice. Despite the uncertainty, you will have to decide.

For example, should you drink less? Numbers cannot answer that question for you. They can seem like the ideal excuse to stop thinking, but they cannot provide quick and easy answers. At best, they will help you navigate the terrain.

And it's not just that the figures are inconclusive; other factors play a role that are not captured in the numbers. How important is drinking alcohol to me? How much risk should I take with my health? How healthy am I, generally speaking? These are things you will have to work out for yourself.

In short, be aware of your feelings, investigate the available information and accept uncertainty. And then make your own decision.

A final tip: watch out for a conflict of interest

In June 2018, another report appeared about a study into the effects of alcohol on our health.[21] This report was not about the results of the study but about the fact that the study had been stopped prematurely. In the experiment, the first of its kind, people had to drink one glass of alcohol every day for six years or, in the control group, none at all.

Previously, there had been commotion over the fact that the American National Institutes of Health had received the better part of a million dollars funding from the alcohol industry. Heineken, Carlsberg and other manufacturers had co-funded the study.[22] And now it emerged from internal research by the NIH that the scientists had promised the alcohol industry that the study could give the 'level of evidence necessary if alcohol is to be recommended as part of a healthy diet'.[23]

The study had been set up in such a way that all the benefits would be visible, while the harmful effects would be overlooked. The length of the trial was too short, because many types of cancer develop slowly. Particular groups of people – those with cancer in the family, for instance – were excluded. All this was done under the pretext of safety, but it also reduced the likelihood that cancer would develop and be linked to alcohol consumption.

If you want to recognise number misuse, it is important to grasp errors of reasoning and to understand your own gut feelings. But maybe the most significant question you should ask is: who's behind the number? Does he or she have a vested interest in the outcome?

AFTERWORD

PUTTING NUMBERS BACK WHERE THEY BELONG

Throughout the years, poor usage of numbers has often driven me to despair. The fallacies in thinking that continue to crop up, the gut feelings that lead to false interpretations, the interests that hold sway over the process of establishing the truth – it's enough to make you lose heart. It's such a shame, because numbers can help us understand the world and can make it a better place. But we should handle them with care. And treat them as critically as we do words.

It's time to put numbers in their place. Since I began writing about numbers, I have come across ever more inspiring initiatives that do just that – criticising misuse of numbers or calling into question their role. Initiatives that show we are not powerless.

Take GDP. Over the past few years, unease has begun to surface about the limitations of GDP and the dominant role it plays in relation to government policy. Various measures that could replace or complement GDP have been suggested. Some countries, for example, now measure their citizens' happiness;[1] the OECD created the Better Life Index, a broader indicator that takes into account factors like the environment or the job market in a particular country;[2] and Statistic Netherlands (CBS) has recently started measuring 'general concept of well-being', which, among other things, studies the effects of our prosperity on future generations.[3]

Political polls have also been at the receiving end of scrutiny. Critics are fed up with fevered speculation over small shifts in single polls that turned them into major news stories. As a result, 'polling aggregators' have flourished, which gather election polls. Such a compilation of results should give a more reliable estimate and – hopefully – cancel out individual polls' biases. Some aggregators take simple averages, like RealClearPolitics, while others, such as FiveThirtyEight, build more elaborate models to come to an estimate.

Problems in science, such as publication bias and p-hacking (the intentional search for significant results), are also starting to be addressed. Since 2012, economists and other social science researchers have been registering their experiments with the American Economic Association before they actually conduct their research.[4] This means it's immediately clear what they are planning to do, so that they do not endlessly look for significant results later on.

For a long time, replications – repeat studies – were unpopular because scientists were expected to come up with juicy, new results, but during the past few years such studies have been appearing with increasing frequency. For example, the American Center for Open Science set up the Reproducibility Project for psychology studies.[5] Two hundred and seventy scientists repeated hundreds of psychological trials and found the studied effect to be smaller than in the original studies and often less significant. Now there are even scientific journals that publish just replications.[6]

But, you may ask, what if you are not a policy-maker or scientist? What can you then do about the dominance of numbers in your life?

Take the education of your children. We hear a great deal about the dominance of test results. But there are teachers and schools who go in the other direction: they award fewer numerical grades. Economics teacher Anton Nanninga, for example, decided to use words rather than numbers to indicate to his pupils how well they had performed. Now he can no longer hide behind a number, he explained in an interview with the NIVOZ Foundation.[7] 'I have to give proper feedback now.' German teacher Martin Ringenaldus no longer gives numerical marks in some of his classes. 'A relief!' he tweeted me. 'Greater motivation amongst the students and a relaxed atmosphere (no test pressure). Even the declensions aren't a problem anymore.'[8] These are only experiments, but they show that the use of numbers is not a given, it's a choice.

Another area in which numbers play an important role is in your job. In the Bijenkorf department store in the Netherlands, targets play a part on the shop floor. In some branches, sales assistants were tasked with asking customers for an evaluation of their performance – preferably including the name of the assistant.[9] Not exactly a reliable measure, as it turned out: a Bijenkorf employee told the Dutch current affairs programme *Nieuwsuur* how her colleagues had asked their entire family to award them a 9 or a 10 to improve their overall score.[10] Also, the scores made the employees stressed out – it was even rumoured that they were used in appraisals. The Bijenkorf was criticised in national media and the Dutch Federation of Trade Unions (FNV) called on customers only to award sales assistants a rating of 10 out of 10. The uproar has helped, as the store changed its policy: customers can still leave their rating, but sales assistants no longer have to ask for personal feedback.

It even appears that there is room for resistance against big data algorithms. Take the OpenSCHUFA initiative.[11] SCHUFA is the largest credit bureau in Germany. Their credit scores have major consequences for the financial situation of an individual, but the company refuses to make their algorithm public. However, according to German law you, as a citizen, can request your own report. In 2018, Open Knowledge Foundation and Algorithm Watch, in fact, called on German citizens to apply for their credit reports and forward them on. With sufficient data, they would be able to reverse engineer the algorithm. Within a few months, more than 25,000 people had requested their own credit report.[12] Each of these people considered it important to understand what was hiding behind these numbers.

All these positive initiatives demonstrate that the dominant role of numbers in our lives is not a given, but something we can resist. Whether you are a journalist or a policy-maker, teacher or doctor, police officer or statistician – numbers affect your life. And you have a right to interfere.

We invented numbers, so it is up to us how we use them.

CHECKLIST:

WHAT TO DO WHEN YOU ENCOUNTER A NUMBER

You encounter a number, say on the news.[1] Do you want to know if it can be trusted? Then ask yourself the following six questions. If you cannot answer the questions because it's impossible to find the right information, then reject the number straightaway. If a researcher has not been clear about his or her methods, it's not worth your attention.

1. Who is the messenger?
Has a politician presented a statistic demonstrating his policy is good for the economy? Have the manufacturers of a particular chocolate bar funded research proving that chocolate is good for your health? Look extra carefully and try to find additional sources.

2. What do I feel?
Does the number make you feel happy, angry or sad? Be careful not to accept it or brush it aside unquestioningly. Be aware of your gut feelings and look for sources with a different perspective.

3. How has it been standardised?
Does the number deal with an invented concept, such as economic growth or intelligence? Pay extra attention. Which choices were made when the measurement was done? Has the number been

blown up into something it is not, like using GDP to describe our general well-being? Try to find research that measures the concept in a different way.

4. How has the data been collected?
The number is probably based on data collected in a study. Imagine you are one of the participants in the research. Do any of the questions push you in a particular direction? Are the circumstances such that you would rather not tell the truth? Then take the number with an extra pinch of salt. And was the sample random or not? If not, remember that the number only applies for the specific group that was studied.

5. How has the data been analysed?
Does the number relate to an alleged causal link? Ask the following three questions: could the link have come about by chance? Do other factors play a part? Could the causal link work the other way around? In any case, never take one study as the gospel truth. Look for meta-studies that show what the entire research field says. Or look for a compilation of polls, like those compiled by the polling website FiveThirtyEight.

6. How have the numbers been presented?
Finally, a few things to look out for in the presentation of numbers.

- **An average**: if there are outliers that can hike the average up or pull it down, then the figure does not say much about the common situation.
- **A precise figure**: there are all kinds of reasons why figures cannot be altogether 100 per cent precise. Do not allow yourself to be taken in by bogus precision.

- **A ranking**: adjacent places in a ranking frequently do not indicate a significant difference between the two because there are margins of error.
- **A risk**: it's useless to know that there is an x per cent greater chance that you may get a particular disease, if you do not know what the percentage is *of*. If this chance is small to begin with, then an increase of x per cent will also be small.
- **A graph**: an odd vertical axis can distort the results. Watch out that it isn't stretched, or squashed together.

NOTES

FOREWORD: CAPTIVATED BY NUMBERS

1. I wrote about my encounter with Juanita earlier on my blog *Out of the Blauw* and on the Oikocredit Nederland (Oikocredit Netherlands) blog. I have not been able to contact her to show her this story, which is why I have given her a pseudonym.

CHAPTER 1: NUMBERS CAN SAVE LIVES

1. For the story about Florence Nightingale I used Mark Bostridge's biography *Florence Nightingale – The Woman and Her Legend* (Viking, 2008) and the article 'Florence Nightingale Was Born 197 Years Ago, and Her Infographics Were Better Than Most of the Internet's' by Cara Giaimo which appeared on 12 May 2017 in *Atlas Obscura*.
2. Florence Nightingale, *Notes on Matters Affecting the Health, Efficiency, and Hospital Administration of the British Army* (Harrison and Sons, London, 1858). She used data that had been collected by British and French statisticians. This can be found in 'Florence Nightingale, Statistics and the Crimean War' by Lynn McDonald, *Statistics in Society* (May 2013).
3. Hugh Small, 'Florence Nightingale's Hockey Stick', *Royal Statistical Society* (7 October 2010).
4. Iris Veysey, 'A Statistical Campaign: Florence Nightingale and Harriet Martineau's England and her Soldiers', *Science Museum Group Journal* (3 May 2016).
5. Harold Raugh, *The Victorians at War, 1815–1914: An Encyclopedia of British Military History* (ABC-CLIO, 2004).
6. Lynn McDonald, *Florence Nightingale and Hospital Reform: Collected Works of Florence* (Wilfrid Laurier University Press, 2012), page 442.

7. Hugh Small, 'Florence Nightingale's Statistical Diagrams', presentation to a Research Conference organised by the Florence Nightingale Museum, 18 March 1998.
8. This has been the case since 1811 at the Registry of Births, Deaths and Marriages. The system had been introduced in some regions in France as early as 1796.
9. Ian Hacking, 'Biopower and the Avalanche of Printed Numbers', *Humanities in Society* (1982).
10. Meg Leta Ambrose, 'Lessons from the Avalanche of Numbers: Big Data in Historical Perspective', *Journal of Law and Policy for the Information Society* (2015).
11. For this paragraph I have drawn on *Sapiens* by Yuval Noah Harari (Harvill Secker, London, 2014).
12. For this paragraph I have used *Seeing Like a State* by James Scott (Yale University Press, New Haven, 1998).
13. Ken Alder, 'A Revolution to Measure: The Political Economy of the Metric System in France', in *Values of Precision* (Princeton University Press, 1995), pp. 39–71.
14. James Scott, *Seeing Like a State* (Yale University Press, New Haven, 1998).
15. Ken Alder, 'A Revolution to Measure: The Political Economy of the Metric System in France', in *Values of Precision* (Princeton University Press, 1995), pp. 39–71.
16. This remark was inspired by James Scott, who writes in *Seeing Like a State* (Yale University Press, New Haven,1998): 'For centralizing elites, the universal meter was to older, particularistic measurement practices as a national language was to the existing welter of dialects.'
17. Mars Climate Orbiter Mishap Investigation Board, *Phase I Report* (10 November 1999).
18. It was the time of the Enlightenment and the 'scientific revolution', during which scientists founded their thinking and research on reason and universal principles.
19. 'Appendix G: Weights and Measures', *CIA World Factbook* (consulted on 26 July 2018).
20. Meg Leta Ambrose, 'Lessons from the Avalanche of Numbers: Big Data in Historical Perspective', *Journal of Law and Policy for the Information Society* (2015).
21. This can be found in 'Biopower and the Avalanche of Printed Numbers', *Humanities in Society* (1982). In this article, Hacking also describes the list of diseases William Farr drew up with his colleagues.
22. This remark was inspired by Yuval Noah Harari, who wrote the following about

our numbering system in *Sapiens* (Harvill Secker, London, 2014): 'It has become the world's dominant language'.
23. Hans Nissen, Peter Damerow and Robert Englund, *Archaic Bookkeeping: Early Writing and Techniques of Economic Administration in the ancient Near East* (University of Chicago Press, 1994).
24. 'Census', *Wikipedia* (consulted on 26 July 2018).
25. Jelke Bethlehem, 'The Rise of Survey Sampling', Statistics Netherlands (2009).
26. Ian Hacking, in 'Biopower and the Avalanche of Printed Numbers', *Humanities in Society* (1982), called the growth during this period 'exponential'. The remainder of this paragraph was also based on Hacking's article.
27. 'General Register Office', *Wikipedia* (consulted on 28 July 2018).
28. Ian Hacking, 'Biopower and the Avalanche of Printed Numbers', *Humanities in Society* (1982).
29. My thoughts about Adolphe Quetelet have been based on *The End of Average* by Todd Rose, in a Dutch version titled *De mythe van het gemiddelde*, translated by Theo van der Ster and Aad Markenstein (Bruna Uitgevers, 2016).
30. Nightingale called Quetelet 'the creator of statistics' in a letter she wrote to him. Gustav Jahoda, 'Quetelet and the Emergence of the Behavioral Sciences', *SpringerPlus* (2015).
31. This revolution was to lead to Belgium's independence from the Netherlands.
32. Quetelet not only saw the 'average man' as a statistical phenomenon, but also as an idealised image of humankind.
33. Stephen Stigler, 'Darwin, Galton and the Statistical Enlightenment', *Journal of the Royal Statistical Society* (2010).
34. I came across Archibald Cochrane in *Superforecasting* by Philip Tetlock and Dan Gardner (Random House Books, 2016). I have based this paragraph on Cochrane's autobiography *One Man's Medicine* (BMJ Books, London, 1989), which he co-wrote with Max Blythe.
35. Marcus White, 'James Lind: The Man who Helped to Cure Scurvy with Lemons', BBC News (4 October 2016). We now know that citrus fruits contain vitamin C, which can prevent or combat scurvy.
36. 'Nutritional yeast', *Wikipedia* (consulted on 26 July 2018).
37. In his autobiography, Cochrane does not clarify which consequences he meant.
38. I base this description on Archie Cochrane's autobiography, *One Man's Medicine* (BMJ Books, London, 1989). The anecdote also appears in *Superforecasting* by Philip Tetlock and Dan Gardner (Random House Books, 2016).
39. David Isaacs, 'Seven Alternatives to Evidence Based Medicine', *BMJ* (18 December 1999).

40. This is also called 'cognitive dissonance'.
41. This experiment is described in *Ending Medical Reversal* by Vinayak Prasad and Adam Cifu, (Johns Hopkins University Press, Baltimore, 2015). In an earlier article, these researchers looked at all the articles which had been published over a ten-year period in one scientific journal. They came up with a shocking result: in almost 140 cases the accepted methods turned out not to work. (Prasad et al., 'A Decade of Reversal: An Analysis of 146 Contradicted Medical Practices', *Mayo Clinical Proceedings*, 18 July 2013.)
42. Sanne Blauw, 'Vijf woorden die volgens statistici de wereld kunnen redden', ('Five Words which Statisticians Believe Can Save the World') *De Correspondent* (10 February 2017).
43. Anushka Asthana, 'Boris Johnson Left Isolated as Row Grows over £350m Post-Brexit Claim', *Guardian* (17 September 2017).
44. 'Called to Account', *The Economist* (3 September 2016).

CHAPTER 2: THE DUMB DISCUSSION ABOUT IQ AND SKIN COLOUR

1. For the history of the IQ-test in this chapter I made grateful use of *The Mismeasure of Man*, by Stephen Jay Gould, in a Dutch version translated by Ton Maas and Frits Smeets (Uitgeverij Contact, Amsterdam, 1996). In later research, aspects of Gould's book have been called into question, but not his account of the IQ-test. If you want to read more about the discussion, I recommend reading Jason Lewis, David DeGusta, Marc Meyer, Janet Monge, Alan Mann and Ralph Holloway, 'The Mismeasure of Science: Stephen Jay Gould versus Samuel George Morton on Skulls and Bias', *PLoS Biology* (7 June 2011), and also Michael Weisberg and Diane Paul, 'Morton, Gould, and Bias: A Comment on "The Mismeasure of Science"', *PloS* Biology (19 April 2016).
2. E.G. Boring, Yerkes' assistent, selected 160,000 cases and analysed the figures.
3. Jeroen Pen, '"Racisme? Het gaat op de arbeidsmarkt om IQ"' ('"Racism? IQ is what Counts in the Job Market"'), *Brandpunt+* (9 June 2016).
4. For this paragraph I used Gavin Evans 'The Unwelcome Revival of "Race Science"', *Guardian* (2 March 2018).
5. Margalit Fox, 'Arthur R. Jensen Dies at 89; Set Off Debate About I.Q.', *New York Times* (1 November 2012).
6. Richard Herrnstein and Charles Murray, *The Bell Curve* (Free Press, 1994).
7. Nicholas Wade, *A Troublesome Inheritance* (Penguin, London, 2014). Some

140 geneticists wrote a letter to protest against Wade's statements, see 'Letters: "A Troublesome Inheritance"', *New York Times* (8 August 2014).
8. D.J. Kevles, 'Testing the army's intelligence: Psychologists and the military in World War I', *Journal of American History* (1968).
9. Discrimination through quotas was done in a subtle way: the quota was set at 2 per cent of the number of immigrants from that country already resident. The data from the 1890 census was used, which featured relatively few Southern and Eastern Europeans, instead of the data from the most recent census in 1920.
10. Six million, Allan Chase estimates in *The Legacy of Malthus* (Knopf, New York, 1977). Chase assumes that immigration remained unchanged compared to before 1924.
11. Andrea DenHoed, 'The Forgotten Lessons of the American Eugenics Movement', *New Yorker* (27 April 2016).
12. The figures are taken from William Dickens and James Flynn, 'Black Americans Reduce the Racial IQ Gap: Evidence from Standardization Samples' *Psychological Science* (2006). I use the test results from the Wechsler Adult Intelligence Scale from the year 1995.
13. Malcolm Gladwell, 'None of the Above', *New Yorker* (17 December 2007).
14. David Reich, 'How Genetics Is Changing Our Understanding of Race', *New York Times* (23 March 2018).
15. D'Vera Cohn, 'Millions of Americans Changed their Racial or Ethnic Identity from One Census to the Next', *Pew Research Center*, 5 May 2014.
16. In order to measure IQ-scores the test is taken among a representative sample and then recalculated in such a way that they fall within a 'normal distribution' with an average of a 100 points and so that 68 per cent of the people score between 85 and 115.
17. 'Inkomens van personen (Individual Income)', *cbs.nl* (consulted on 6 September 2018).
18. Binet's story is related in Stephen Jay Gould, *The Mismeasure of* Man, in its Dutch version translated by Ton Maas and Frits Smeets, (Uitgeverij Contact, Amsterdam, 1996), pp. 195–204.
19. This description of money and other invented concepts was inspired by *Sapiens* by Yuval Noah Harari (Harvill Secker, London, 2014).
20. I base my account of the history of GDP on *GDP: A Brief but Affectionate History* by Diane Coyle (Princeton University Press, 2014).
21. Although Kuznets is often seen as the inventor of GDP, he built on methods already in existence, for example those created by British statistician Colin Clark.

22. Simon Kuznets, 'National Income, 1929–1932', *National Bureau of Economic Research* (7 June 1934).
23. Strictly speaking it was not GDP, but 'Gross National Product' (GNP). GDP is the value of goods and services within a particular country, while GNP measures the value of goods and services produced by the inhabitants of that country (so even if these services are actually carried out outside the country's borders).
24. Dutch Prime Minister Mark Rutte, for instance, introduced tax increases and cuts in order to stimulate the economy and, in so doing, exit the recession. According to the Netherlands Bureau for Economic Policy Analysis, the country is in a recession when GDP has shrunk for a minimum of two quarters.
25. This intermezzo is based on my article 'Hoe precieze cijfers ons misleiden and de geschiedenis bepalen' ('How Precise Figures Mislead Us and Determine History'), *De Correspondent* (1 December 2015).
26. Enrico Berkes and Samuel Williamson, 'Vintage Does Matter, The Impact and Interpretation of Post War Revisions in the Official Estimates of GDP for the United Kingdom', measuringworth.com (consulted on 15 August 2018). It's worth noting that newer datasets were produced annually, which showed differences compared to the previous year.
27. Shane Legg and Marcus Hutter, 'A collection of definitions of intelligence', *Frontiers in Artificial Intelligence and Applications* (2007).
28. 'Wechsler Adult Intelligence Scale', *Wikipedia* (consulted on 30 July 2018).
29. I came across Luria's story in a TED Talk by James Flynn, 'Why Our IQ Levels Are Higher than Our Grandparents" (March 2013). The account of Luria's travels to Uzbekistan can be found in his autobiography, *The Autobiography of Alexander Luria: A Dialogue with The Making of Mind*, co-written with Michael Cole and Karl Levitin (Psychology Press, 1979, republished in 2010).
30. These examples have been inspired by a speech about GDP by Bobby Kennedy on 18 March 1968.
31. Anne Roeters, *Een week in kaart (A Week Charted)*, the Netherlands Institute for Social Research (Sociaal and Cultureel Planbureau, 2017).
32. Tucker Higgins, 'Trump Suggests Economy Could Grow at 8 Or 9 Percent If He Cuts the Trade Deficit', *CNBC* (27 July 2018).
33. The budget deficit can be no more than 3 per cent of GDP and the national debt cannot exceed 60 per cent of GDP. It's easier for a country to meet these requirements when it has a higher GDP.
34. Many traineeships in business and civil service feature assessments which include an IQ test or comparable evaluations.
35. I base my story about Spearman on *The Mismeasure of Man* by Stephen Jay

Gould, in its Dutch version, translated by Ton Maas and Frits Smeets (Uitgeverij Contact, 1996).
36. He used the 'factor analysis' method, in which a mountain of numbers is simplified into common 'factors'. Spearman concluded that just one factor could explain many of the differences between children.
37. Stephen Jay Gould, *The Mismeasure of Man*, in its Dutch version, translated by Ton Maas and Frits Smeets (Uitgeverij Contact, 1996).
38. Charles Spearman, 'General Intelligence Objectively Measured and Determined', *The American Journal of Psychology* (April 1904).
39. Edwin Boring, 'Intelligence as the Tests Test It', *New Republic* (1923).
40. The *Landelijk Kader Nederlandse Politie 2003–2006 (National Dutch Police Structural Plan 2003–2006)* featured fine quota for the different police forces. In later agreements between government and the police the requirements for the number of fines had been removed, but police forces continued to use production quota. Fine Quota were finally banned by Ivo Opstelten (VVD Liberal Party, Justice and Security). I wrote about Fine Quota before, in the article 'Hoe cijferdictatuur het werk van leraren, agenten and artsen onmogelijk maakt' ('How the Dictatorship of Numbers Makes the Work of Teachers, Police Officers and Doctors Intolerable'), which I published together with Jesse Frederik on *De Correspondent* (5 January 2016).
41. Peter Campbell, Adrian Boyle and Ian Higginson, 'Should We Scrap the Target of a Maximum Four Hour Wait in Emergency Departments?', *BMJ* (2017).
42. This wording of Goodhart's Law comes from '"Improving Ratings": Audit in the British University System' by Marilyn Strathern, *European Review* (July 1997). Charles Goodhart first articulated his idea in two articles from 1975. For further details, see 'Goodhart's Law: Its Origins, Meaning and Implications for Monetary Policy' by K. Alec Chrystal and Paul Mizen in *Central Banking, Monetary Theory and Practice* (Edward Elgar Publishing, 2003).
43. Stephen Jay Gould, *The Mismeasure of Man*, in its Dutch version, translated by Ton Maas and Frits Smeets (1996).
44. Kevin McGrew, 'The Cattell–Horn–Carroll Theory of Cognitive Abilities', in *Contemporary Intellectual Assessment: Theories, Tests, and Issues* (The Guilford Press, 1996).
45. This paragraph is based on *GDP: A Brief but Affectionate History* by Diane Coyle (Princeton University Press, 2014).
46. He won the 'Nobel Memorial Prize in Economic Sciences'. Strictly speaking this is not a Nobel Prize, but it is often referred to as such.
47. *Human Development Report 2019*, United Nations Development Programme

(2019). With these kind of figures it's important to remember that they contain a margin of error, a concept that is covered in Chapter 3. This means that the scores from some countries cannot be differentiated statistically, because the data always contains some 'noise'.
48. *Jinek*, KRO-NCRV (21 December 2017).
49. Maarten Back, 'AD publiceert alleen nog de 75 beste olliebollenkramen' ('AD only Publishes the 75 Best Doughnut Stalls'), *NRC* (22 December 2017).
50. Herm Joosten, 'Voor patiënten is de AD ziekenhuis-lijst (vrijwel) zinloos' ('The AD Hospital Table is (Virtually) Useless for Patients'), *de Volkskrant* (10 October 2014).
51. Sometimes moral choices lurk without creators realising it. Economist Martin Ravallion studied the HDI and found a strange result: a country with a reduced life expectancy could still find itself achieving a higher HDI by growing just a small amount in terms of income. Because the different dimensions were grouped in one number, they had become interchangeable. When Ravallion began calculating, he came to an absurd conclusion: a human life was worth less in one country than in another. The absolute rock-bottom was Zimbabwe, where an extra year of life equated fifty euro cents. In rich countries, on the other hand, the price rose to 8,000 euros or more. See Martin Ravallion, 'Troubling Tradeoffs in the Human Development Index', *Journal of Development Economics* (November 2012).
52. I wrote earlier about the definition of hunger in 'Waarom we veel minder weten van ontwikkelingslanden dan we denken' ('Why We Know Much Less about Developing Countries than We Think'), *De Correspondent* (30 June 2015).
53. *The State of Food Insecurity in the World*, Food and Agriculture Organization (2012).
54. James Flynn, 'Why Our IQ Levels Are Higher than Our Grandparents', TED.com (March 2013).
55. Earlier researchers had spotted something in some samples, but James Flynn was the first to study it structurally.
56. In some countries you also see an 'anti-Flynn-effect', reductions in IQ. Data from Norwegian men showed that their IQ had dropped between 1975 and 1990. See Bernt Bratsberg and Ole Rogeberg, 'Flynn Effect and Its Reversal Are Both Environmentally Caused', *PNAS* (26 June 2018).
57. Yerkes used the term 'moron' for educationally subnormal, a term that is only used as a term of abuse these days.
58. Carl Brigham, *A Study of American Intelligence* (Princeton University Press, 1923).
59. While he was studying philosophy, someone had told him he would never be a true philosopher. 'Never!' he wrote in 1909. 'What a momentous word. Some

recent thinkers seem to have given their moral support to these deplorable verdicts by affirming that an individual's intelligence is a fixed quantity, a quantity that cannot be increased. We must protest and react against this brutal pessimism; we must try to demonstrate that it is founded upon nothing!' See Gould, pages 183–184.

60. Diane Coyle, *GDP: A Brief but Affectionate History* (Princeton University Press, 2014).
61. Malcolm Gladwell, 'None of the above', *New Yorker* (17 December 2007). Gladwell's Italics.
62. Anandi Mani, Sendhil Mullainathan, Eldar Shafir and Jiaying Zhao, 'Poverty Impedes Cognitive Function', *Science* (30 August 2013).
63. Tamara Daley, Shannon Whaley, Marian Sigman, Michael Espinosa and Charlotte Neumann, 'IQ On the Rise: The Flynn Effect in Rural Kenyan Children', *Psychological Science* (May 2003).
64. William Dickens and James Flynn, 'Black Americans Reduce the Racial IQ Gap: Evidence from Standardization Samples', *Psychological Science* (2006).
65. Angela Hanks, Danyelle Solomon, Christian Weller, *Systematic Inequality: How America's Structural Racism Helped Create the Black-White Wealth Gap*, Center for American Progress (21 February 2018).
66. Alana Semuels, 'Good School, Rich School; Bad School, Poor School', *The Atlantic* (25 August 2016); Alvin Chang, 'Living in a Poor Neighborhood Changes Everything about Your Life', Vox.com (4 April 2018).
67. Marianne Bertrand and Esther Duflo, 'Field Experiments on Discrimination', in *Handbook of Field Experiments* (Elsevier, 2017).

CHAPTER 3. WHAT A SHADY SEX STUDY SAYS ABOUT SAMPLING

1. Truman was already President, because he had taken over the position after the death of Franklin D. Roosevelt.
2. The newspaper relied on the judgement of its political correspondent Arthur Sears Henning, who had predicted the elections using polls and other information. See also 'The Untold Story of "Dewey Defeats Truman"' by Craig Silverman, *Huffington Post* (5 December 2008).
3. Michael Barbaro, 'How Did the Media – How Did We – Get This Wrong?', *New York Times* (9 November 2016).
4. To be more precise, Wang stated that he would eat an insect if Trump won more than 240 seats in the electoral college; Trump won 290. See Sam Wang,

'Sound Bites and Bug Bites', *Princeton Election Consortium* (4 November 2016). Wang posted the tweet on 19 October 2016.
5. Alexandra King, 'Poll Expert Eats Bug on CNN After Trump Win', *CNN* (12 November 2016).
6. Jelke Bethlehem, 'The Rise of Survey Sampling', Statistics Netherlands (2009).
7. Tom Smith, 'The First Straw? A Study of the Origins of Election Polls', *Public Opinion Quarterly* (1990).
8. Smith argued that the elections of 1824 were the 'first seriously contested' since 1800. After 1800 changes had been introduced into the system, which meant that the elections would be decided primarily by a popular majority.
9. Sarah Igo, *The Averaged American: Surveys, Citizens and the Making of a Mass Public* (Harvard University Press, Cambridge, Mass., 2007).
10. This was not the first time that cracks were appearing in the image of polls. In 1936, the magazine *Literary Digest* – up until then an authority in the field – had predicted that Alf Landon would win. He lost. *Literary Digest* had to fold a year later.
11. Alfred Kinsey, Wardell Pomeroy and Clyde Martin, *Sexual Behavior in the Human Male* (W.B. Saunders Company, 1948).
12. Frederick Mosteller, *The Pleasures of Statistics: The Autobiography of Frederick Mosteller* (Springer, 2010).
13. David Spiegelhalter, *Sex by Numbers* (Profile Books, London, 2015).
14. Thomas Rueb, 'Eén op de tien wereldburgers is homoseksueel' ('One in Ten People in the World is Gay'), nrc.nl (24 July 2012).
15. Sarah Igo, *The Averaged American: Surveys, Citizens and the Making of a Mass Public* (Harvard University Press, Cambridge, Mass., 2007).
16. For my discussion of Kinsey's research and the account of the three statisticians in this chapter I used the following three books: James Jones, *Alfred C. Kinsey: A Life* (Norton, New York, 1997); Sarah Igo, *The Averaged American: Surveys, Citizens and the Making of a Mass Public* (Harvard University Press, Cambridge, Mass., 2007); David Spiegelhalter, *Sex by Numbers* (Profile Books, London, 2015).
17. Kinsey argued in his report that 100,000 observations would ultimately be needed. He hoped to publish a more extended version of his study, but it never happened.
18. 'The Kinsey Interview Kit', *The Kinsey Institute for Research in Sex, Gender and Reproduction* (1985).
19. The italics in this quote are mine.
20. David Spiegelhalter, *Sex by Numbers* (Profile Books, London, 2015).

21. These figures have been taken from the Natsal-3-Study and are mentioned in Chapter 3 in David Spiegelhalter, *Sex by Numbers* (Profile Books, London, 2015).
22. Michele Alexander and Terri Fisher, 'Truth and consequences: Using the bogus pipeline to examine sex differences in self-reported sexuality', *Journal of Sex Research* (2003). The experiment is disucssed in Chapter 3 in David Spiegelhalter, *Sex by Numbers* (Profile Books, London, 2015). The 2.6 bed partners were observed in a group in which there was a likelihood that another student was looking on as well. There was another research group, in which the respondents were in a closed room; in this group the average number of bed partners was 3.4.
23. Guy Harling, Dumile Gumede, Tinofa Mutevedzi, Nuala McGrath, Janet Seeley, Deenan Pillay, Till W. Bärnighausen and Abraham J. Herbst, 'The Impact of Self-Interviews on Response Patterns for Sensitive Topics: A Randomized Trial of Electronic Delivery Methods for a Sexual Behaviour Questionnaire in Rural South Africa', *BMC Medical Research Methodology* (2017).
24. I came across this poll in the BBC Radio 4 programme *More or Less*, which covered the poll on 5 December 2017. The criticism I express here and in the following section, is discussed there as well. Tim Harford, the programme's presenter, spoke to Prithwiraj Mukherjee, who wrote under the handle @peelaraja on Twitter: 'If you are in my marketing research class and design such a survey I will fail you' (21 November 2016).
25. Jelke Bethlehem, 'Terrorisme een groot probleem? Het is maar net hoe je het vraagt' ('Is Terrorism a Big Problem? It Depends How You Frame the Question'), *peilingpraktijken.nl* (2 October 2014).
26. David Spiegelhalter, *Sex by Numbers* (Profile Books, London, 2015).
27. Page 6 of the report states that the number of black men taking part in the study was too small to be able to say anything about them.
28. 'Internet Users per 100 Inhabitants', *unstats.un.org* (consulted on 31 July 2018).
29. Jeffrey Arnett, 'The Neglected 95%: Why American Psychology Needs to Become Less American', *American Psychologist* (October 2008).
30. Joseph Henrich, Steven Heine and Ara Norenzayan, 'The Weirdest People in the World?', *Behavioral and Brain Sciences* (June 2010).
31. A possible explanation for this is that people in modern societies have got used to square angles, such as those found in buildings or urban squares. This has taught our brain a particular visual trick, which turns out to be a problem in the Müller-Lyer illusion.
32. These and the next paragraphs have been based on the book *Inferior* by Angela Saini (HarperCollins Publishers, 2018).
33. 'Drug Safety: Most Drugs Withdrawn in Recent Years Had Greater Health

Risks for Women', United States Government Accountability Office (19 January 2001).
34. Archibald Cochrane and Max Blythe, *One Man's Medicine* (BMJ Books, London, 1989).
35. Dana Carney, Amy Cuddy and Andy Yap, 'Power Posing: Brief Nonverbal Displays Affect Neuroendocrine Levels and Risk Tolerance', *Psychological Science* (2010).
36. Eva Ranehill, Anna Dreber, Magnus Johannesson, Susanne Leiberg, Sunhae Sul and Roberto Weber, 'Assessing the Robustness of Power Posing: No Effect on Hormones and Risk Tolerance in a Large Sample of Men and Women', *Psychological Science* (2015). In 2018, together with two colleagues, Cuddy presented a study showing that the expansive pose did indeed have positive effects, but when the data was analysed anew by different researchers once again no proof for the effect of the powerful posture materialised. See Marcus Crede, 'A Negative Effect of a Contractive Pose Is Not Evidence for the Positive Effect of an Expansive Pose: Commentary on Cuddy, Schultz, and Fosse (2018)', unpublished manuscript, available on *SSRN* (12 July 2018).
37. Katherine Button, John Ioannidis, Claire Mokrysz, Brian Nosek, Jonathan Flint, Emma Robinson and Marcus Munafò, 'Power failure: why small sample size undermines the reliability of neuroscience', *Nature Reviews: Neuroscience* (May 2013).
38. This anecdote was described in Sarah Igo, *The Averaged American: Surveys, Citizens and the Making of a Mass Public* (Harvard University Press, Cambridge, Mass., 2007).
39. Perhaps you'll have noticed the figure of 18,000 does not match the 11,000 cases in the two reports. Kinsey and his colleagues interviewed 18,000 people, but not every observation ended up in the reports, for example, those of black men or of people who were interviewed after the publication of the reports.
40. A technical point: an unrepresentative cross-section of the population may, thanks to chance, still emerge; but because you know the chance of this happening while randomising, you can quantify the degree of representativeness.
41. This was reported in 'Kinsey', an episode in the documentary series *American Experience*, first broadcast on 14 February 2015.
42. Richard Pérez-Peña, '1 in 4 Women Experience Sex Assault on Campus', *New York Times* (21 September 2015). I found out about this poll through an article on the Huffington Post by Brian Earp: '1 in 4 Women: How the Latest Sexual Assault Statistics Were Turned into Click Bait by the *New York Times*' (28 September 2015).

43. David Cantor, Reanne Townsend and Hanyu Sun, 'Methodology Report for the AAU Campus Climate Survey on Sexual Assault and Sexual Misconduct', *Westat* (12 April 2016).
44. The calculations are as follows. If the remaining 80 per cent is victim: 0,2*0.25+0.8*1=0.85 (85 per cent). If the other 80 per cent is not a victim: 0,2*0.25+0.8*0=0.05 (5 per cent).
45. Such a bandwith takes into account non-response and it is assumed that the sample is representative and the questions have been asked correctly.
46. Go to *https://goodcalculators.com/margin-of-error-calculator/* and enter 'Population Size'; this is the group you are interested in. In this case: American men, during Kinsey's time the population totalled sixty million. In this (hypothetical) example, the 'Sample Size' is equal to 100 and the 'Proportion Percentage' was 50 per cent. The margin of error that emerges is 9.8 per cent, so the percentage could have been as low as 40.2 per cent and as high as 59.8 per cent. (These are the intervals for 95 per cent reliability.)
47. David Weigel, 'State Pollsters, Pummeled by 2016, Analyze What Went Wrong', *Washington Post* (30 December 2016).
48. Because America uses the electoral college system, the person who wins the *popular vote* is not necessarily the winner of presidential elections.
49. I chose ABC News/ *Washington Post* because it was awarded an A+ by *FiveThirtyEight*, the highest ranking the data website gives to a pollster. The 4 per cent margin of error is mentioned in Scott Clement and Dan Balz, '*Washington Post* – ABC News Poll: Clinton Holds Four-Point Lead in Aftermath of Trump Tape', *Washington Post* (16 October 2016).
50. Nate Silver, 'The Real Story of 2016', *fivethirtyeight.com* (19 January 2017).
51. 'NOS Nederland Kiest: De Uitslagen' ('The Netherlands Goes To the Polls, the Results'), *NOS* (18 March 2015). Stax made the comment on 2:07:50.
52. James Jones, *Alfred C. Kinsey: A Life* (Norton, 1997).
53. John Bancroft, 'Alfred Kinsey's Work 50 Years on', in a new edition of *Sexual Behavior in the Human Female* (Indiana University Press, 1998).
54. Mr X is what Jones calls the man in his biography on Kinsey.
55. This quote comes from James Jones, *Alfred C. Kinsey: A Life* (Norton, 1997), as do other quotes in the following paragraphs.

CHAPTER 4: SMOKING CAUSES LUNG CANCER (BUT STORKS DO NOT DELIVER BABIES)

1. For my discussion of the tobacco industry in this chapter I use: Robert Proctor,

Golden Holocaust: Origins of the Cigarette Catastrophe and the Case for Abolition (University of California Press, 2011); Naomi Oreskes and Erik Conway, *Merchants of Doubt: How a Handful of Scientists Obscured the Truth on Issues from Tobacco Smoke to Global Warming* (Bloomsbury, 2012); and Tim Harford, 'Cigarettes, Damn Cigarettes and Statistics', *Financial Times* (10 April 2015).

2. Ernest Wynder, Evarts Graham and Adele Croninger, 'Experimental Production of Carcinoma with Cigarette Tar', *Cancer Research* (December 1953).
3. 'Background Material on the Cigarette Industry Client', a memo from 15 December 1953, which can be found in the Industry Documents Library, a collection of documents from the tobacco industry.
4. With the exception of Ligget & Myers, which preferred to ignore this entire enterprise.
5. 'A Frank Statement to Cigarette Smokers', 4 January 1954.
6. Naomi Oreskes and Erik Conway, *Merchants of Doubt* (Bloomsbury, London, 2012), page 15.
7. Darrell Huff, *How to Lie with Statistics* (Victor Gollancz, 1954). I used the Penguin edition from 1991.
8. J. Michael Steele, 'Darrell Huff and Fifty Years of *How to Lie with Statistics*', *Statistical Science*, Institute of Mathematical Statistics (2005).
9. 'NUcheckt: Helpt gin-tonic tegen hooikoorts?' ('NU checks: Is Gin and Tonic Good For Hayfever?'), *NU.nl* (3 May 2018).
10. Anouk Broersma, 'Wegscheren schaamhaar vergroot kans op soa' ('Shaving Pubic Hair Increases Your Chances of Getting an STD'), *de Volkskrant* (6 December 2016).
11. Liesbeth De Corte, 'Chocolade is wél gezond, maar enkel en alleen de pure variant' ('Chocolate is Healthy, But Only in the Dark Variety'), *AD* (5 May 2018).
12. Sumner Petroc, Vivian-Griffiths Solveiga, Boivin Jacky, Williams Andy, Venetis Christos A, Davies Aimée et al. 'The association between exaggeration in health related science news and academic press releases: retrospective observational study', *BMJ* (10 December 2014).
13. Jonathan Schoenfeld and John Ioannidis, 'Is Everything We Eat Associated with Cancer? A Systematic Cookbook Review', *American Journal of Clinical Nutrition* (January 2013).
14. I also discuss Paul in 'Deze statistische fout wordt in bijna elk debat gemaakt (en zo pik je haar eruit)' ('This Statistical Mistake is Made in Almost Every Debate (And this is the Way to Spot it)'), *De Correspondent* (8 March 2016).
15. Lotto Odds https://www.lottery.co.uk/lotto/odds (last checked on January 10th 2020).

16. *www.tylervigen.com/spurious-correlations* (consulted on 3 August 2018).
17. Randall Munroe, 'Significant', *xkcd.com*.
18. Brian Wansink, David Just and Collin Payne, 'Can Branding Improve School Lunches?', *Archives of Pediatrics and Adolescent Medicine* (October 2012).
19. Brian Wansink and Koert van Ittersum, 'Portion Size Me: Plate-Size Induced Consumption Norms and Win-Win Solutions for Reducing Food Intake and Waste', *Journal of Experimental Psychology: Applied* (December 2013).
20. Stephanie Lee, 'Here's How Cornell Scientist Brian Wansink Turned Shoddy Data into Viral Studies about How We Eat', *BuzzFeed News* (25 February 2018).
21. Archibald Cochrane and Max Blythe, *One Man's Medicine* (BMJ Books, London,1989).
22. I wrote about this study in 'Deze statistische fout wordt in bijna elk debat gemaakt (en zo pik je haar eruit)' ('This statistical mistake is made in almost every debate (And this is the way to spot it)'), *De Correspondent* (8 March 2016).
23. 'Borstsparende therapie bij vroege borstkanker leidt tot betere overleving' ('Lumpectomy in Early Breast Cancer Leads to Better Survical Chances') *IKNL* (10 December 2015).
24. For an overview of the reporting, see 'Is borstsparend opereren en bestralen beter dan amputeren?' ('Is a Lumpectomy Combined with Radiotheray Better than a Mastectomy?'), *Borstkankervereniging Nederland (Netherlands Breast Cancer Association)* (15 December 2015).
25. Marissa van Maaren, Linda de Munck, Luc Strobbe and Sabine Siesling, 'Toelichting op berichtgeving over onderzoek naar borstkankeroperaties' ('Comments on Reporting on Studies into Breast Cancer Surgery'), *IKNL* (17 December 2015).
26. Ronald Veldhuizen, 'Zijn borstamputaties tóch gevaarlijker dan borstsparende operaties?' ('Are Mastectomies More Dangerous than Lumpectomies after all?'), *de Volkskrant* (17 December 2015).
27. Here, too, a third factor could play a part: smoking. Smokers tend to be slimmer and have worse survival chances. Andrew Stokes and Samuel Preston, 'Smoking and Reverse Causation Create an Obesity Paradox in Cardiovascular Disease', *Obesity* (2015).
28. This chapter looks primarily at lung cancer and not at different adverse health effects such as other types of cancer and heart failure.
29. I talked earlier about this news item in my TEDx Talk, 'How to Defend Yourself against Misleading Statistics in the News', *TEDx Talks* (3 November 2016).
30. 'Moeten we misschien iets minder vlees eten?' ('Should we Eat a Little Less Meat?'), *Zondag met Lubach (Sunday with Lubach)*, VPRO (1 November 2015).
31. Martijn Katan, 'NRC Opinie 29-10-2015: Vleeswaren en darmkanker' ('NRC

Opinion 29-10-2015: Processed Meats and Bowel Cancer'), *mkatan.nl* (29 October 2015).
32. 'Q&A on the Carcinogenicity of the Consumption of Red Meat and Processed Meat', World Health Organization (October 2015).
33. Fritz Lickint, 'Tabak und Tabakrauch als ätiologischer Faktor des Carcinoms' ('Tobacco and tobacco smoke as aetiological factor of carcimoa'), *Zeitschrift for Krebsforschung und klinische Onkologie (Journal of Cancer Research and Clinical Oncology)* (December 1930).
34. Richard Doll and Austin Bradford Hill, 'A Study of the Aetiology of Carcinoma of the Lung', *British Medical Journal* (1952).
35. Robert Proctor, *Golden Holocaust: Origins of the Cigarette Catastrophe and the Case for Abolition* (University of California Press, 2011).
36. The tobacco industry has been compelled to release documents. You are able to view all the material on the website of *Legacy Tobacco Documents Library*.
37. 'The only #climatechange chart you need to see http://natl.re/wPKpro (h/t @PowelineUS)', @NationalReview on Twitter, 14 December 2015.
38. Roz Pidcock, 'How Do Scientists Measure Global Temperature', *CarbonBrief* (16 January 2015).
39. 'GISS Surface Temperature Analysis', *data.giss.nasa.gov* (consulted on 8 January 2018).
40. Roz Pidcock, 'Scientists Compare Climate Change Impacts at 1.5C and 2C', *CarbonBrief* (21 April 2016).
41. This is a 'moving average', which means that it is calculated for a period of five years, which moves a year at a time.
42. 'Statement by Darrell Huff', *Truth Tobacco Industry Document*.
43. Ronald Fisher, *Smoking. The Cancer Controversy: Some Attempts to Assess the Evidence* (F.R.S. Oliver and Boyd, 1959).
44. David Salsburg, *The Lady Tasting Tea* (A.W.H. Freeman, 2001).
45. David Roberts, 'The 2 Key Points Climate Skeptics Miss', *Vox.com* (11 December 2015).
46. Claude Teague, 'Survey of Cancer Research' (1953).
47. 'WHO Statement on Philip Morris Funded Foundation for a Smoke-Free World', World Health Organization (28 September 2017).
48. Naomi Oreskes and Erik Conway, *Merchants of Doubt: How a Handful of Scientists Obscured the Truth on Issues from Tobacco Smoke to Global Warming* (Bloomsbury, London, 2012).
49. Martijn Katan, 'Hoe melkvet gezond wordt' ('How Milk Fat Becomes Healthy'), *mkatan.nl* (30 January 2010).

50. Christie Aschwanden, 'There's No Such Thing As "Sound Science"', *FiveThirtyEight* (6 December 2017).
51. Personal communication with David Daube's son, mentioned in Robert Proctor, *Golden Holocaust: Origins of the Cigarette Catastrophe and the Case for Abolition* (University of California Press, 2011).
52. Alex Reinhart, 'Huff and Puff', *Significance* (October 2014).

CHAPTER 5: WE SHOULD NOT BE TOO FIXATED ON NUMBERS IN THE FUTURE

1. The story about Jenipher comes from a TED Talk by Shivani Siroya: 'A Smart Loan for People with No Credit History (Yet)', *TED.com* (February 2016).
2. For this chapter, I made grateful use of *Weapons of Math Destruction* by Cathy O'Neil (Crown, 2016).
3. Sean Trainor, 'The Long, Twisted History of Your Credit Score', *Time* (22 July 2015).
4. Numbers also play a part in facial recognition, as it involves measuring someone's face.
5. 'Data Never Sleeps 5.0', domo.com (consulted on 14 August 2018).
6. Brian Resnick, 'How Data Scientists Are Using AI for Suicide Prevention', *Vox.com* (9 June 2018).
7. Celine Herweijer, '8 Ways AI Can Help Save the Planet', *World Economic Forum* (24 January 2018).
8. 'No Longer Science Fiction, AI and Robotics Are Transforming Healthcare', *PWC Global* (consulted on 15 August 2018).
9. Mallory Soldner, 'Your Company's Data Could End World Hunger', *TED.com* (September 2016).
10. Louise Fresco, 'Zeg me wat u koopt en ik zeg wat u stemt' ('Tell Me What You Buy and I will Tell You How You Vote'), *NRC* (16 November 2016).
11. Marc Hijink, 'Hoe bepaalt de verzekeraar hoe veilig jij rijdt?' ('How Does Your Insurer Decide How Safe Your Driving is?'), *NRC* (5 April 2018).
12. Maurits Martijn, 'Baas Belastingdienst over big data: "Mijn missie is gedragsverandering"' ('Tax Authorities Chief: "My Mission Is Behavioural Change"'), *De Correspondent* (21 April 2015).
13. Julia Dressel and Hany Farid, 'The Accuracy, Fairness, and Limits of Predicting Recidivism', *ScienceAdvances* (17 January 2018).
14. Brian Christian and Tom Griffiths, *Algorithms to Live by* (Henry Holt and Company, 2016).

15. Cathy O'Neil, *Weapons of Math Destruction* (Crown, 2016).
16. In 1959, computer scientist Arthur Samuel coined the term *machine learning*, using the following definition: 'field of study that gives computers the ability to learn without being explicitly programmed'.
17. 'Our Story', zestfinance.com (consulted on 14 August 2018).
18. 'Zest Automated Machine Learning', zestfinance.com (consulted on 14 August 2018).
19. For this paragraph I made use of 'U staat op een zwarte lijst' ('You Have Been Blacklisted') by Karlijn Kuijpers, Thomas Muntz and Tim Staal, *De Groene Amsterdammer* (25 October 2017).
20. Julia Dressel and Hany Farid, 'The Accuracy, Fairness and Limits of Predicting Recidivism', *ScienceAdvances* (17 January 2018).
21. 'Background Checking – The Use of Credit Background Checks in Hiring Decisions', *Society for Human Resource Management* (19 July 2012). In theory, you can refuse permission for a check. But you have little choice: by refusing, you may throw away your chances of a job.
22. Amy Traub, *Discredited*, Demos (February 2013).
23. 'Credit Reports', *Last Week Tonight with John Oliver*, HBO (10 April 2016).
24. In the survey mentioned previously, 45 per cent of the employers cited as a justification that they wanted to prevent criminality, 19 per cent to assess the candidate's reliability.
25. Jeremy Bernerth, Shannon Taylor, H. Jack Walker and Daniel Whitman, 'An Empirical Investigation of Dispositional Antecedents and Performance-Related Outcomes of Credit Scores', *Journal of Applied Psychology* (2012).
26. Kristle Cortés, Andrew Glover and Murat Tasci, 'The Unintended Consequences of Employer Credit Check Bans on Labor and Credit Markets', Working Paper no. 16-25R2, Federal Reserve Bank of Cleveland (January 2018).
27. Sean Illing, 'Proof That Americans Are Lying About Their Sexual Desires', *Vox.com* (2 January 2018).
28. Seth Stephens-Davidowitz, *Everybody Lies* (Bloomsbury Publishing, London, 2017).
29. '*All data is credit data*', Douglas Merrill says in his TEDx Talk 'New credit scores in a new world: Serving the Underbanked' (13 April 2012).
30. Karlijn Kuijpers, Thomas Muntz and Tim Staal, 'U staat op een zwarte lijst' ('You Have Been Blacklisted'), *De Groene Amsterdammer* (25 October 2017).
31. *Report to Congress Under Section 319 of the Fair and Accurate Credit Transactions Act of 2003*, Federal Trade Commission (December 2012).
32. Lauren Brennan, Mando Watson, Robert Klaber and Tagore Charles, 'The

Importance of Knowing Context of Hospital Episode Statistics When Reconfiguring the NHS', *BMJ* (2012).
33. Jim Finkle and Aparajita Saxena, 'Equifax Profit Beats Street View as Breach Costs Climb', *Reuters* (1 March 2018).
34. Cathy O'Neil, *Weapons of Math Destruction* (Crown, 2016).
35. 'Stat Oil', *Economist* (9 February 2013).
36. Ron Lieber, 'American Express Kept a (Very) Watchful Eye on Charges', *New York Times* (30 January 2009).
37. Robinson Meyer, 'Facebook's New Patent, "Digital Redlining", and Financial Justice' *The Atlantic* (25 September 2015).
38. 'Stat Oil', *Economist* (9 February 2013).
39. Chris Anderson, 'The End of Theory', *Wired* (23 June 2008).
40. Jesse Frederik, 'In de economie valt een appel niét altijd naar beneden (ook al zeggen economen vaak van wel)' ('In the Economy, the Apple does not Always Fall to the Ground (Even though Economists say it Does)'), *De Correspondent* (24 September 2015).
41. Erick Schonfeld, 'Eric Schmidt Tells Charlie Rose Google is "Unlikely" to Buy Twitter and Wants to Turn Phones into TVs', *TechCrunch* (7 March 2009).
42. To be more precise: the algorithm was supposed to predict the number of doctor visits. See David Lazer, Ryan Kennedy, Gary King and Alessandro Vespignani, 'The Parable of Google Flu: Traps in Big Data Analysis', *Science* (14 March 2014). I have also used this article in the subsequent paragraphs.
43. This correlation is not completely accidental, because the high school basketball season runs more or less concurrently with the flu season.
44. For my account of this experiment I use: Tim Harford, *The Logic of Life* (Random House, 2009); and Roland Fryer, Jacob Goeree and Charles Holt, 'Experience-Based Discrimination: Classroom Games', *The Journal of Economic Education* (Spring 2005).
45. 'Planning Outline for the Construction of a Social Credit System (2014–2020)', translated into English by Rogier Creemers, *China Copyright and Media* (14 June 2014). The subsequent quote is also from this document.
46. Rogier Creemers, 'China's Social Credit System: An Evolving Practice of Control', *SSRN* (9 May 2018).
47. Alipay website, *intl.alipay.com* (consulted on 15 August 2018).
48. For this and the following paragraphs I have used: Rachel Botsman, 'Big Data Meets Big Brother as China Moves to Rate Its Citizens', *Wired* (21 October 2017); Mara Hvistendahl, 'Inside China's Vast New Experiment in Social Ranking', *Wired* (14 December 2017).

49. Paul Lewis, '"Fiction is Outperforming Reality": How YouTube's Algorithm Distorts the Truth', *Guardian* (2 February 2018).
50. 'FTC Report Confirms Credit Reports Are Accurate', *CISION PR Newswire* (11 February 2013).
51. Maurits Martijn and Dimitri Tokmetzis, *Je hebt wél iets te verbergen*, ('You Do Have Something to Hide'), *De Correspondent* (2016).

CHAPTER 6: OUR PSYCHOLOGY DECIDES THE VALUE OF NUMBERS

1. 'Een glas alcohol is eigenlijk al te veel' ('One Glass of Alcohol is One Too Many), *nos.nl* (13 April 2018).
2. A reworked version of this chapter appeared on *De Correspondent* with the title 'Waarom slimme mensen domme dingen zeggen' ('Why Clever People Say Stupid Things') on 18 July 2018. Parts of this chapter have been inspired by Tim Harford, 'Your Handy Postcard-Sized Guide to Statistics', timharford.com, published previously in *Financial Times* (8 February 2018).
3. Angela Wood et al, 'Risk Thresholds for Alcohol Consumption: Combined Analysis of Individual-Participant Data for 599 912 Current Drinkers in 83 Prospective Studies', *The Lancet* (14 April 2018).
4. @VinayPrasadMD on Twitter (28 April 2018).
5. 'Skills Matter: Further Results from the Survey of Adult Skills' (OECD Publishing, 2016).
6. 'PISA 2012 Results: Ready to Learn Students' Engagement, Drive and Self-Beliefs (Volume III)' (OECD Publishing, 2013).
7. Sanne Blauw, 'Waarom we slechte cijfers zoveel aandacht geven' ('Why We Pay So Much Attention to Bad Numbers'), *De Correspondent* (15 June 2017).
8. Sanne Blauw, 'Het twaalfde gebod: wees je bewust van je eigen vooroordelen' ('The Twelfth Commandment: Be Aware of Your Own Prejudices'), *De Correspondent* (24 February 2016).
9. Dan Kahan, Ellen Peters, Erica Cantrell Dawson and Paul Slovic, 'Motivated Numeracy and Enlightened Self-Government', *Behavioural Public Policy* (May 2017). In the discussion of this study I have made grateful use of Ezra Klein, 'How Politics Makes Us Stupid', *Vox.com* (6 April 2014).
10. Respondents were asked for their party political preference and ideology. Kahan and colleagues translated this, in line with scientific literature, but into a divide into 'liberal Democrats' and 'conservative Republicans'.
11. The findings have often been replicated, not only by Kahan and colleagues, but

also by others. For examples, see Dan Kahan, Asheley Landrum, Katie Carpenter, Laura Helft and Kathleen Hall Jamieson, 'Science Curiosity and Political Information Processing', *Advances in Political Psychology* (2017).
12. Beth Kowitt, 'The Paradox of American Farmers and Climate Change', *fortune. com* (29 June 2016).
13. Ezra Klein, 'How Politics Makes Us Stupid', *Vox.com* (6 April 2014).
14. "'Een extra glas alcohol kan je leven met 30 minuten verkorten'" ('One Extra Glass of Alcohol Can Shorten Your Life by 30 Minutes'), *AD* (13 April 2018).
15. Dan Kahan, Asheley Landrum, Katie Carpenter, Laura Helft and Kathleen Hall Jamieson 'Science Curiosity and Political Information Processing', *Advances in Political Psychology* (2017). In my discussion of the study I make grateful use of Brian Resnick, 'There May Be an Antidote to Politically Motivated Reasoning. And It's Wonderfully Simple', *Vox.com* (7 February 2017).
16. In the remainder of this chapter I refer to science curiosity as 'curiosity'.
17. Tim Harford, 'Your Handy Postcard-Sized Guide to Statistics', timharford.com, published previously in *Financial Times* (8 February 2018).
18. 'Animal Models in Alcohol Research', *Alcohol Alert* (April 1994).
19. Chiara Scoccianti, Béatrice Lauby-Secretan, Pierre-Yves Bello, Véronique Chajes and Isabelle Romieu, 'Female Breast Cancer and Alcohol Consumption: A Review of the Literature', *American Journal of Preventive Medicine* (2014).
20. *Richtlijnen goede voeding 2015 (Guidelines for Healthy Eating)*, Netherlands Health Council (2015).
21. Roni Caryn Rabin, 'Major Study of Drinking Will Be Shut Down', *New York Times* (15 June 2018).
22. Roni Caryn Rabin, 'Federal Agency Courted Alcohol Industry to Fund Study on Benefits of Moderate Drinking', *New York Times* (17 March 2018).
23. Owen Dyer, '$100m Alcohol Study Is Cancelled amid Pro-Industry "Bias"', *BMJ* (19 June 2018).

AFTERWORD: PUTTING NUMBERS BACK WHERE THEY BELONG

1. Sanne Blauw, 'Waarom je beter geluk dan rendement kunt meten' ('Why It's Better to Measure Happiness than Financial Returns'), *De Correspondent* (20 March 2015).
2. 'OECD Better Life Index', http://www.oecdbetterlifeindex.org (consulted on 17 August 2018).
3. *Monitor brede welvaart 2018 (Monitor of Well-being: a Broader Picture)*, Netherlands Statistics (2018).

4. 'AEA RCT Registry', http://www.socialscienceregistry.org (consulted on 16 August 2018). Registered Reports from the Center for Open Science is another example.
5. 'Estimating the Reproducibility of Psychological Science', Open Science Collaboration, *Science* (2015).
6. See for example the *International Journal for Re-Views in Empirical Economics*.
7. Geert Bors, 'Leraar zijn in relatie (2): je bent je eigen instrument' (Being a teacher in relation (2): You Are Your Own Agent), *Stichting NIVOZ* (4 July 2018).
8. 'I've been teaching for 3 years now [in secondary vocational school] without grading the students. A relief! Greater motivation amongst the students and a relaxed atmosphere (no test pressure). Even the declensions aren't a problem. Very proud of the little rascals. Am the only one to work like this in school, 'though. The primary classes want to introduce it as well.', @bijlesduits on Twitter, 30 May 2018.
9. Sheila Sitalsing, 'Dappere verkoopsters van de Bijenkorf bewijzen: protesteren tegen onzin heeft zin' ('Brave Bijenkorf Department Store Sales Assistants Prove: Protesting Against Nonsense is Useful'), *de Volkskrant* (22 May 2018).
10. 'Steeds meer beoordelingen: "Dit geeft alleen maar stress"' ('More and More Evaluations Only Lead to Stress"), *Nieuwsuur* (24 April 2018).
11. http://www.openschufa.de (consulted on 17 August 2018).
12. selbstauskunft.net/schufa. Consulted on 18 September; at that point, 27,959 applications had been made.

CHECKLIST: WHAT TO DO WHEN YOU ENCOUNTER A NUMBER

1. The six questions in this checklist have been inspired by similar lists, such as *Your Handy Postcard-Sized Guide to Statistics* by Tim Harford, the last chapter of *How to Lie with Statistics* by Darrell Huff and *The Pocket Guide to Bullshit Prevention* by Michelle Nijhuis.

FURTHER READING

Parts of this book appeared earlier on the *Correspondent*, on my blog 'Out of the Blauw', and on the Oikocredit Nederland blog.

My hope is that this book is accessible to everyone, which is why I have kept it compact and have – through necessity – not been able to go more deeply into some topics. Happily, quite a few wonderful books have already been written about the misuse of statistics, the history of our numbered society and other topics that I have covered.

How to Lie with Statistics remains a must-read, despite Darrell Huff's dubious history. I can also recommend *Proofiness* by Charles Seife and *How Not to Be Wrong* by Jordan Ellenberg. In order to keep up with statistical misuse in current affairs, listen to *More or Less* on BBC Radio 4, follow Nate Silver's website FiveThirtyEight, and keep an eye out for the fact-check columns in newspapers.

If you want to know more about the history of our numbered society, I suggest you read *Seeing Like a State* by James C. Scott and *Sapiens* by Yuval Noah Harari. For the history of the IQ test, see *The Mismeasure of Man* by Stephen Jay Gould. Diane Coyle writes beautifully about GDP in her *GDP: A Brief But Affectionate History*. For a historical view on polls, *The Averaged American* by Sarah Igo is a good start, and for more information about research into sex, *Sex By Numbers* by David Spiegelhalter is an absolute must. The tobacco industry's practices have been chronicled in *Golden Holocaust* by Robert Proctor, and in *Merchants of Doubt* by Naomi Oreskes and Erik Conway. To read more about big data algorithms, go to *Weapons of Math Destruction* by Cathy O'Neil and *Hello World* by Hannah Fry. The psychological processes needed for interpreting numbers have been described sublimely by Daniel Kahneman in *Thinking Fast and Slow*. *Superforecasting*, by Philip Tetlock and Dan Gardner, shows how our psyche co-determines how we make predictions and interpret reality.

Finally, I thoroughly enjoyed reading the following biographies: *One Man's Medicine* by Archibald Cochrane and Max Blythe, *Florence Nightingale* by Mark Bostridge, and *Alfred C. Kinsey*, by James Jones.

ACKNOWLEDGEMENTS

A book is more than a collection of pages. Writing is more than typing as many words as you can. And even though my name is on the cover of this book, it's the product of many people around me. It's sometimes said that it takes a village to raise a child. In the case of this book, a medium-sized provincial town might be a better analogy.

First, I would like to thank all online subscribers to the *Correspondent*. Over the years you have presented me with ideas, sharpened my thoughts and given me the confidence that this topic deserved a title. What good fortune to be able to spend my working days in such warm and inquisitive company.

I found similar warmth and curiosity at the Netherlands Institute for Advanced Study where, as journalist-in-residence, I was allowed to work on my book for five months. Thanks to the other fellows and NIAS staff, I managed to take the big plunge needed to write this book. Huge thanks to the Fonds Bijzondere Journalistieke Projecten (Extraordinary Journalistic Projects Fund), which enabled me to have this experience.

Following an appeal in my newsletter, dozens of readers offered to proofread chapters. I was overwhelmed by the response. I would like to thank Berend Alberts, Gerard Alberts, Lotte van Dillen, Eefje Dons, Marcel Haas, Eva de Hullu, Jenneke Krüger, Anke Richters,

Judith ter Schure, Eduard van Valkenburg and Joris van Vugt for their extremely useful comments.

Many thanks also to Casper Albers, Anna Alberts, Jelke Bethlehem, Rogier Creemers, Ninette van Hasselt, Wanda de Kanter, Daniël Lakens, Tom Louwerse, Marijke van Mourik and Daniel Mügge, who cast their expert eye over the manuscript.

This book has received endorsements from Barbara Baarsma, Rutger Bregman, Pieter Derks, José van Dijck, Femke Halsema, Bas Haring, Rosanne Hertzberger and Ionica Smeets. I find it quite extraordinary that you decided to make room in your busy schedules to read it. Thank you.

On to my colleagues at the *Correspondent*. Only a few years ago, I only knew you in print; now, you have become flesh-and-blood people. You are much more than simply work. Thank you for your support and your company.

I would like to thank Rob Wijnberg for coming up with the Dutch title of this book and for creating my dream job. Many thanks also to Dimitri Tokmetzis for taking a critical look over my manuscript; to Maite Vermeulen, who has taught me so much about journalism and who has become a much-loved friend; and to Rutger Bregman, friend and mentor. With great meticulousness, Annelieke Tillema combed the errors from the text. And Veerle van Wijk helped enormously with tying up loose ends.

I also owe a lot of gratitude to 'team international'. Rebecca Carter at Janklow & Nesbit, and Juliet Brooke and Louise Court at Sceptre – thank you for believing in this book even before being able to read it. And thanks to Suzanne Heukensfeldt Jansen, and everyone who helped with the English version.

My biggest thanks go to the 'hard core'. Andreas Jonkers, thank you for your sharp comments and incessant enthusiasm to get this book noted.

Milou Klein Lankhorst, I was just a novice when we first began talking about this book. Thank you for your faith; it's an honour to work with you.

And Harminke Medendorp, you were so often perched on my shoulder during the lonely writing hours. You taught me lessons that will stay with me for the rest of my writer's life. And what an amazing person you are.

This book would never have seen the light of day without my great love, Middelburg, the town I grew up in. It sounds so romantic, a writer's retreat, but if my family and friends had not dragged me out from under my noise-cancelling headphones every now and again, I would have gone crazy.

Anna de Bruyckere, Carlotta van Hellenberg Hubar and Carlijn Janssen, what a joy to have you in my life for so many years now. Thank you for your humour, faith and patient ear.

Hylke Blauw and Marieke Langen, your family is a bright ray of sunshine in my life. Just tell Mies, Pia and Pepijn that Auntie Sannie will come and babysit again soon.

Jurre Blauw and Jetje Blauw-Lindo, thank you for asking me to do something even more scary than writing a book. The day I had the honour of officiating at your wedding was one of the most beautiful of my life.

Tjeerd Blauw and Dominique Willemse, I am extremely grateful for all the lunches we had together in Middelburg. I promise I will soon think up an excuse to join you on a daily basis again.

Marijke van Mourik. Mama. This book is dedicated to you. You taught me how to live. Thank you.